AQA GCSE 9-1 Revision

Combined Science

Trilogy

Combined Science

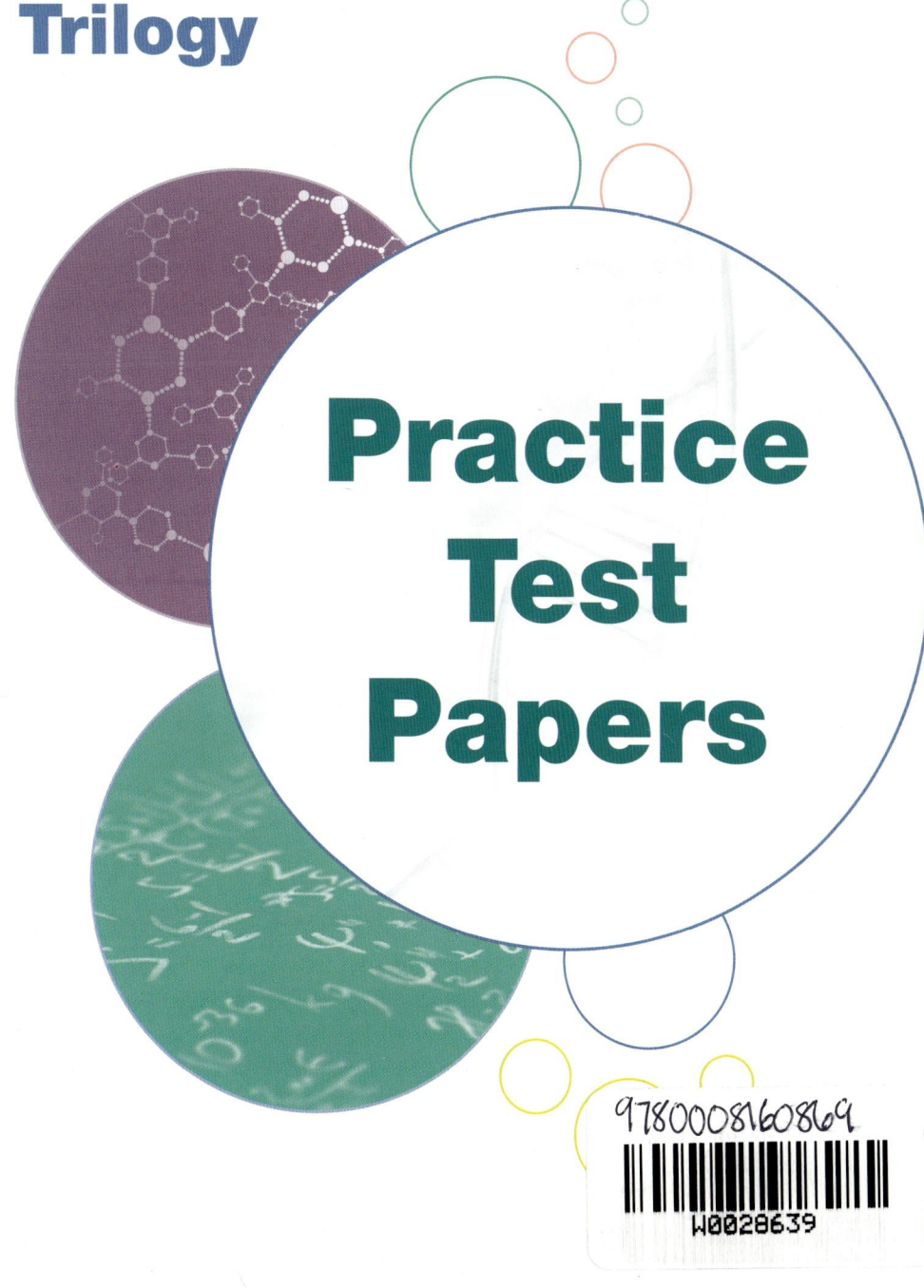

Practice Test Papers

2 x tests

✓ School pack

GCSE 9-1 Revision

	Revision Guide Improve your results with three times more practice opportunities than other revision guides	**All-in-One Revision & Practice** Everything in one place including a revision guide, a workbook and complete test papers	**Grade Boosters** Maximise your marks to get the grade you want
Biology	 9780008160678	 9780008160746	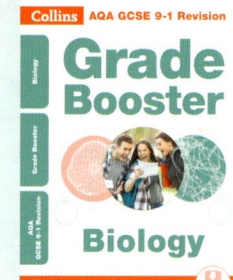 9780008276812
Physics	 9780008160692	 9780008160739	 9780008276836
Chemistry	 9780008160685	 9780008160753	 9780008276829
Combined Science Trilogy	 9780008160791	 9780008160852  9780008160869	 9780008276843

5 EASY WAYS TO ORDER

1. Available from www.collins.co.uk
2. Fax your order to 01484 665736
3. Phone us on 0844 576 8126
4. Email us at education@harpercollins.co.uk
5. Post your order to: Collins Education, FREEPOST RTKB-SGZT-ZYJL, Honley HD9 6QZ

AQA

GCSE

Combined Science: Trilogy   H

SET B – Chemistry: Paper 3 Higher Tier

Author: Paul Lewis

Time allowed: 1 hour 15 minutes

Materials

For this paper you must have:
- a ruler
- a calculator
- the Periodic Table (found at the end of the paper).

Instructions

- Answer **all** questions in the spaces provided.
- Do all rough work in this book. Cross through any work you do not want to be marked.

Information

- There are 70 marks available on this paper.
- The marks for questions are shown in brackets.
- You are expected to use a calculator where appropriate.
- You are reminded of the need for good English and clear presentation in your answers.
- When answering questions 02.2 and 05.4 you need to make sure that your answer:
  – is clear, logical, sensibly structured
  – fully meets the requirements of the question
  – shows that each separate point or step supports the overall answer.

Advice

- In all calculations, show clearly how you work out your answer.

Name: _____

01

01.1 Define the term **exothermic reaction**.

[1 mark]

01.2 Ethan is testing reactions between different chemicals that could be used in a hand warmer.

Table 1.1 shows his results.

Table 1.1

Reaction	Starting temperature (°C)	Final temperature (°C)	Temperature change (°C)
A	18	20	+2
B	19	14	
C	19	29	+10
D	20	48	+28

Calculate the temperature change for reaction B.

Temperature change for reaction B = ... °C

[2 marks]

01.3 What is the **dependent variable** in Ethan's investigations?

[1 mark]

01.4 On the graph paper below, draw a **bar chart** for Ethan's results.

[4 marks]

01.5 When plotting results, a line graph or a bar chart might be used.

Explain why a bar chart was the most appropriate in this case.

_____

_____

[1 mark]

Question 1 continues on the next page

01.6 Using **Table 1.1** and your bar chart, suggest which reaction is **most suitable** for use in a hand warmer.

Explain why you have suggested that reaction.

Investigation ....

Explanation ....

....

[2 marks]

02 **Figure 2.1** shows the structure of **graphene**, a substance taken from graphite.

Figure 2.1

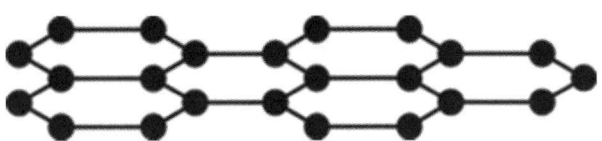

Graphene, graphite and diamond are all made up of atoms of the same element.

It is said that graphene:

- is the strongest material ever measured
- is one of the best conducting materials known
- may have the highest melting point in nature.

02.1 What single element are graphene, graphite and diamond all made from?

....

[1 mark]

Question 2 continues on the next page

02.2 Explain why **graphite** can conduct electricity, is very strong and has a high melting point.

[6 marks]

03 Electrolysis can be used to separate some substances into their components.

03.1 Which **two** of these substances could be separated into their components using electrolysis?

Tick **two** boxes.

Solid carbon dioxide ☐

Molten magnesium chloride ☐

A solution of potassium iodide ☐

Molten sulfur dioxide ☐

Solid potassium iodide ☐

[1 mark]

03.2 The electrolysis of sodium chloride solution (brine) can be carried out within a school laboratory or on a larger scale in industry (**Figure 3.1**).

Figure 3.1

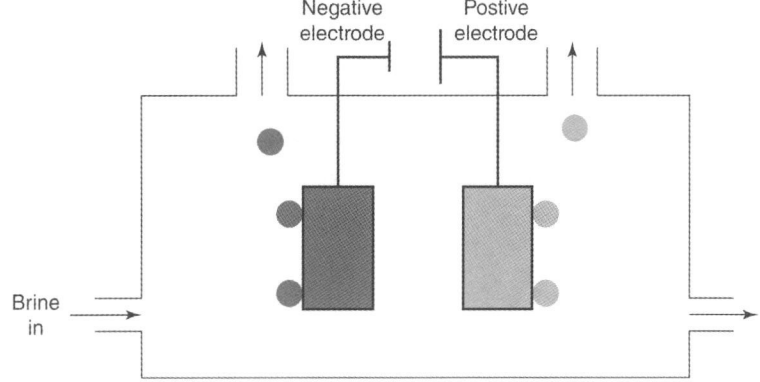

Complete the table to identify the name of the product formed at the positive electrode.

Location	Product
positive electrode	
negative electrode	hydrogen
left in solution	sodium hydroxide

[1 mark]

Question 3 continues on the next page

03.3 Explain why hydrogen ions move to the negative electrode.

[1 mark]

03.4 Explain why hydrogen forms at the negative electrode, but sodium does not.

[1 mark]

03.5 Describe how a hydrogen ion turns into a hydrogen atom.

[1 mark]

04

04.1 A magnesium atom is represented by the symbol:

$$^{24}_{12}Mg$$

Explain why the **mass number** of magnesium is 24.

....

....

....

....

[3 marks]

04.2 Magnesium (Mg) reacts with oxygen (O$_2$) to form magnesium oxide (MgO).

Which type of structure and bonding does magnesium oxide have?

Tick **one** box.

Giant covalent ☐

Giant ionic ☐

Giant metallic ☐

Molecular covalent ☐

[1 mark]

04.3 Explain why magnesium oxide has a high melting point.

....

....

....

[2 marks]

Question 4 continues on the next page

04.4 The equation for the reaction between magnesium and oxygen is:

$$2Mg + O_2 \longrightarrow 2MgO$$

Calculate the minimum mass of magnesium oxide that will be made from 14.75 g of magnesium.

Relative atomic masses: O = 16; Mg = 24

Show your working.

Give your answer to 2 decimal places.

[3 marks]

04.5 In an experiment, the mass balance used to measure the 14.75 g of magnesium had an uncertainty of 5×10^{-3} g

What was the maximum possible mass of magnesium?

[1 mark]

04.6 In the formation of magnesium oxide it is best to use excess oxygen.

Explain why excess oxygen is used.

[2 marks]

05

05.1 Halogens consist of small molecules, and each molecule is made from two atoms of the element.

Figure 5.1 shows the electron structure for an atom of fluorine.

Figure 5.1

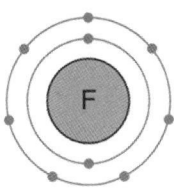

Draw a dot and cross diagram for a **molecule** of fluorine.

[1 mark]

Question 5 continues on the next page

05.2 Three common isotopes of bromine are:

$^{79}_{35}Br$ $^{80}_{35}Br$ $^{81}_{35}Br$

Explain why these three forms of bromine are called 'isotopes'.

[2 marks]

05.3 A sample of bromine consists of:

- two stable isotopes, with mass numbers 79 and 81, and
- one unstable isotope, with mass number 80.

Table 5.1 shows the abundances of the three isotopes.

Table 5.1

Bromine	Relative abundance (%)
Bromine-79	50
Bromine-80	3
Bromine-81	47

Calculate the relative atomic mass of the bromine sample.

Use the data in Table 5.1.

Give your answer to 3 significant figures.

[3 marks]

05.4 The reactivity of the halogens changes down Group 7, and the reactivity of the alkali metals changes down Group 1.

Describe how the reactivity changes down both Group 7 **and** Group 1.

Explain how the reactivity changes down **both** of these groups.

[6 marks]

06 Magnesium sulfate (MgSO$_4$) is a soluble salt made by reacting excess magnesium with sulfuric acid.

06.1 Outline a method that a student could use to obtain solid magnesium sulfate from the magnesium and sulfuric acid.

[3 marks]

06.2 During the reaction, bubbles of hydrogen gas are given off.

How can the student test whether this gas is hydrogen?

[1 mark]

06.3 Calculate the percentage of sulfur in one mole of magnesium sulfate.

Give your answer to 3 significant figures.

Relative atomic masses: S = 32; O = 16; Mg = 24

Percentage of sulfur in one mole of magnesium sulfate = _____ %

[2 marks]

07

07.1 Draw a reaction profile for an **endothermic** reaction on the axis below.

On your profile, label:

- the activation energy and
- the overall energy change.

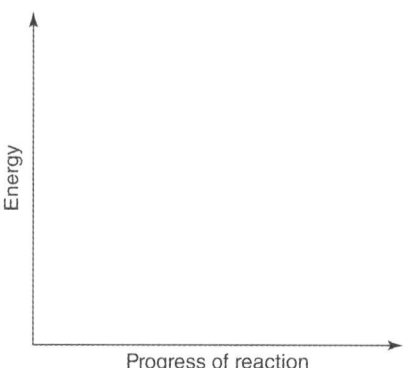

[2 marks]

07.2 **Table 7.1** shows some bond energy data.

Table 7.1

Bond	C–C	C–H	C=O	O=O	O–H
Bond energy (kJ/ mol)	348	413	799	495	463

Calculate the overall energy change for this complete combustion reaction:

$$2C_2H_6 + 7O_2 \longrightarrow 6H_2O + 4CO_2$$

Use the data given in **Table 7.1**.

Overall energy change = ... kJ/mol

[3 marks]

07.3 Explain, in terms of bonds broken and bonds formed, why this reaction is **exothermic**.

[2 marks]

08

08.1 An **atom** of aluminium has the electronic configuration 2,8,3

Draw the electronic structure of an aluminium **ion**.

[2 marks]

08.2 Aluminium is used to manufacture a wide range of products, from drink cans to aircraft.

Use your knowledge of the structure of aluminium to explain why it has properties that are suitable for many different products.

....

....

....

[2 marks]

08.3 Aluminium is manufactured by the electrolysis of a molten mixture of aluminium oxide and cryolite.

Carbon is used for the positive electrode (anode).

Explain why:
- cryolite is added
- a carbon anode is used, and
- the carbon anode eventually needs to be replaced.

[4 marks]

08.4 Aluminium is formed at the negative electrode.

Complete the following half equation.

$$Al^{3+} + \underline{\hspace{2cm}} \longrightarrow \underline{\hspace{2cm}}$$

[1 mark]

08.5 What name is given to the type of reaction that aluminium undergoes in this process?

[1 mark]

END OF QUESTIONS

The Periodic Table

1	2												3	4	5	6	7	0 or 8
																		4 **He** helium 2
7 **Li** lithium 3	9 **Be** beryllium 4												11 **B** boron 5	12 **C** carbon 6	14 **N** nitrogen 7	16 **O** oxygen 8	19 **F** fluorine 9	20 **Ne** neon 10
23 **Na** sodium 11	24 **Mg** magnesium 12												27 **Al** aluminium 13	28 **Si** silicon 14	31 **P** phosphorus 15	32 **S** sulfur 16	35.5 **Cl** chlorine 17	40 **Ar** argon 18
39 **K** potassium 19	40 **Ca** calcium 20	45 **Sc** scandium 21	48 **Ti** titanium 22	51 **V** vanadium 23	52 **Cr** chromium 24	55 **Mn** manganese 25	56 **Fe** iron 26	59 **Co** cobalt 27	59 **Ni** nickel 28	63.5 **Cu** copper 29	65 **Zn** zinc 30		70 **Ga** gallium 31	73 **Ge** germanium 32	75 **As** arsenic 33	79 **Se** selenium 34	80 **Br** bromine 35	84 **Kr** krypton 36
85 **Rb** rubidium 37	88 **Sr** strontium 38	89 **Y** yttrium 39	91 **Zr** zirconium 40	93 **Nb** niobium 41	96 **Mo** molybdenum 42	[98] **Tc** technetium 43	101 **Ru** ruthenium 44	103 **Rh** rhodium 45	106 **Pd** palladium 46	108 **Ag** silver 47	112 **Cd** cadmium 48		115 **In** indium 49	119 **Sn** tin 50	122 **Sb** antimony 51	128 **Te** tellurium 52	127 **I** iodine 53	131 **Xe** xenon 54
133 **Cs** caesium 55	137 **Ba** barium 56	139 **La*** lanthanum 57	178 **Hf** hafnium 72	181 **Ta** tantalum 73	184 **W** tungsten 74	186 **Re** rhenium 75	190 **Os** osmium 76	192 **Ir** iridium 77	195 **Pt** platinum 78	197 **Au** gold 79	201 **Hg** mercury 80		204 **Tl** thallium 81	207 **Pb** lead 82	209 **Bi** bismuth 83	[209] **Po** polonium 84	[210] **At** astatine 85	[222] **Rn** radon 86
[223] **Fr** francium 87	[226] **Ra** radium 88	[227] **Ac*** actinium 89	[261] **Rf** rutherfordium 104	[262] **Db** dubnium 105	[266] **Sg** seaborgium 106	[264] **Bh** bohrium 107	[277] **Hs** hassium 108	[268] **Mt** meitnerium 109	[271] **Ds** darmstadtium 110	[272] **Rg** roentgenium 111	[285] **Cn** copernicium 112		[286] **Uut** ununtrium 113	[289] **Fl** flerovium 114	[289] **Uup** ununpentium 115	[293] **Lv** livermorium 116	[294] **Uus** ununseptium 117	[294] **Uuo** ununoctium 118

Key:
- Metals
- Non-metals

Key labels:
- Relative atomic mass → 1
- Atomic symbol → **H**
- Name → hydrogen
- Atomic/proton number → 1

*The lanthanides (atomic numbers 58–71) and the actinides (atomic numbers 90–103) have been omitted.
The relative atomic masses of copper and chlorine have not been rounded to the nearest whole number.

AQA

GCSE

Combined Science: Trilogy H

SET B – Chemistry: Paper 4 Higher Tier

Author: Paul Lewis

Materials

Time allowed: 1 hour 15 minutes

For this paper you must have:
- a ruler
- a calculator
- the Periodic Table (found at the end of the paper).

Instructions

- Answer **all** questions in the spaces provided.
- Do all rough work in this book. Cross through any work you do not want to be marked.

Information

- There are 70 marks available on this paper.
- The marks for questions are shown in brackets.
- You are expected to use a calculator where appropriate.
- You are reminded of the need for good English and clear presentation in your answers.
- When answering questions 03.3 and 06.1 you need to make sure that your answer:
  - is clear, logical, sensibly structured
  - fully meets the requirements of the question
  - shows that each separate point or step supports the overall answer.

Advice

- In all calculations, show clearly how you work out your answer.

Name: _____

01 Students are investigating the levels of acidity in water samples taken from different parts of the UK.

The students are aiming to find out which area of the UK has the most acidic drinking water by carrying out a neutralisation reaction on samples of water taken from different areas.

The students used the following method:

1. Measure out 30 cm³ of water from a particular area of the UK, using a measuring cylinder.
2. Pour the water into a conical flask.
3. Add 5 drops of indicator.
4. Add 1 molar sodium hydroxide to the water, 1 cm³ at a time, using a pipette, swirling the flask as you add the sodium hydroxide.
5. When the solution turns neutral, stop adding the sodium hydroxide and record the total volume added.
6. Repeat with water from the other areas of the UK.

Table 1.1 shows the results.

Table 1.1

Area of UK	Total volume of sodium hydroxide added (cm³)
Northern Ireland	6
Scotland	4
Wales	12
North West of England	9
South East of England	7

01.1 Use the information in **Table 1.1** to draw a bar chart on the graph paper.

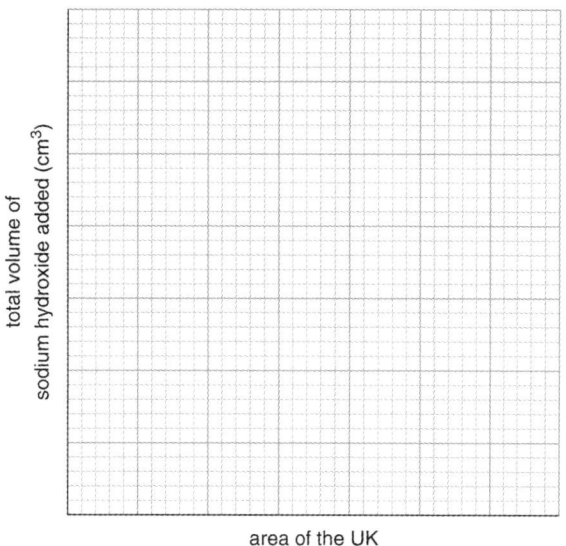

[3 marks]

01.2 Using the information above, identify the independent and dependent variables and one control variable in this investigation.

Independent variable ....

Dependent variable ....

Control variable ....

[3 marks]

Question 1 continues on the next page

01.3 In the method, the students swirled the conical flask when the sodium hydroxide was being added.

State why the students swirled the contents of the conical flask.

....

....

[1 mark]

02 Hydrocarbons, such as alkanes and alkenes, are used as fuels.

02.1 Explain the meaning of the term **hydrocarbon** by completing the sentence below.

A hydrocarbon is a substance made up of only _____ and _____ atoms.

[2 marks]

02.2 For hydrocarbons to be used as fuels, they have to be burnt in oxygen.

What name is given to this type of reaction?

[1 mark]

02.3 These two equations show octane burning in air:

Reaction 1 $2C_8H_{18} + 17O_2 \longrightarrow 16CO + 18H_2O$

Reaction 2 _____ C_8H_{18} + _____ $O_2 \longrightarrow$ _____ CO_2 + _____ H_2O

Balance the equation for **Reaction 2**.

[1 mark]

02.4 What is the name of the carbon compound that is one of the products in **Reaction 1** above?

Suggest why the carbon compound product in **Reaction 1** is different to the carbon compound product in **Reaction 2**.

[2 marks]

Question 2 continues on the next page

02.5 Long hydrocarbons, such as $C_{21}H_{44}$, can be cracked into more useful substances.

Complete this equation to show the cracking of $C_{21}H_{44}$:

$C_{21}H_{44} \longrightarrow C_5H_{12} + C_6H_{12} +$ _____

[1 mark]

02.6 State two conditions needed for cracking.

1. _____

2. _____

[2 marks]

02.7 After cracking, the products form a mixture of alkanes and alkenes.

Describe a chemical test that can be used to identify the **alkenes** produced in such a reaction.

Describe the result of the test.

_____

_____

_____

[2 marks]

03 A student investigated the effect of changing the temperature of an acid on the rate of reaction with magnesium.

Figure 3.1 shows some of the equipment that was available.

Figure 3.1

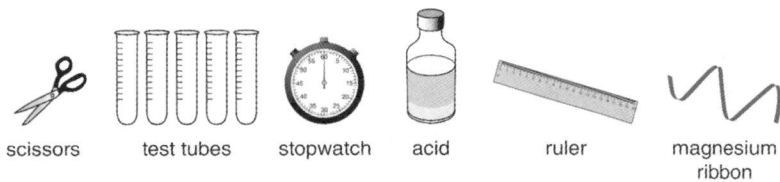

scissors   test tubes   stopwatch   acid   ruler   magnesium ribbon

Table 3.1 shows the student's results.

Table 3.1

Temperature of acid (°C)	Time taken for magnesium to react (s)
20	226
30	179
40	114
50	68
60	23

03.1 Describe any patterns in the results shown in Table 3.1.

[2 marks]

Question 3 continues on the next page

03.2 Explain, in terms of particles and collisions, why a higher temperature affects the rate of reaction.

[3 marks]

03.3 Write a plan for the experiment which the student could have used.

Use the information above, and your own knowledge.

Your answer should include:

- a description of how to make the investigation a fair test
- a description of how to carry out the investigation safely
- a description of any measurements the student needs to take.

[6 marks]

04 Sewage needs to be treated before any water can be released back into the environment.

04.1 Describe the main stages of a sewage treatment system that enables the water to be released back into the environment.

[3 marks]

04.2 What is meant by the term **potable water**?

[1 mark]

04.3 In some areas, the only way to get potable water is to carry out **desalination** of water, such as sea water.

Desalination can involve distillation.

Draw a labelled diagram for the distillation of salt water in a school laboratory.

[3 marks]

04.4 Chlorine is often used to sterilise water.

Describe a test that shows whether chlorine gas is present.

Describe the result of the test.

[2 marks]

05 Crude oil is separated by fractional distillation to give us products such as bitumen, diesel, petrol and LPG.

Figure 5.1 is a diagram of a fractional distillation column.

Figure 5.1

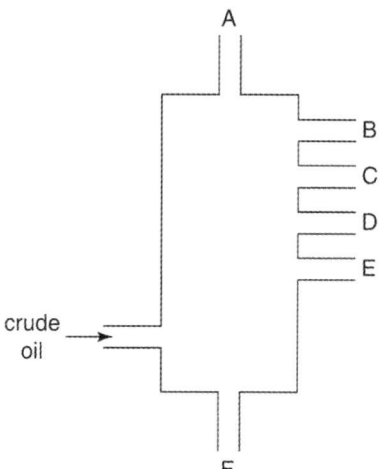

05.1 Describe how fractional distillation is used to separate crude oil.

[4 marks]

05.2 Fractional distillation allows for crude oil to be turned into useful substances.

Table 5.1 shows information about some of the useful fractions that are produced in this process.

Table 5.1

Fraction	Number of carbon atoms in the chain of the molecule	Boiling point (°C)
bitumen	70<	500–700
fuel oil	22–70	350–450
diesel	17–22	220–350
kerosene	10–16	160–220
gasoline	6–10	35–160
LPG	1–5	<35

Describe the trend in the data.

Use **Table 5.1**.

[1 mark]

05.3 Explain why each fraction has a **range** of boiling points and not an exact temperature.

[1 mark]

Question 5 continues on the next page

05.4 When the fuels produced in fractional distillation burn, they often pollute the environment.

Draw one line from each pollutant to the pollution that it causes.

Pollutant	Pollution
Water vapour	Toxic gas that can cause death
Sulfur dioxide	Global dimming
Soot particles	Acid rain
	Greenhouse effect

[2 marks]

06 Read this magazine article, then answer the questions.

> **WASTING RESOURCES**
>
> Many products we have come to rely on are running out as we use up our finite resources. This has led to numerous government initiatives to save our resources, such as the 5p charge for a plastic bag.
>
> The charge has been in place in England since 2015. When it was introduced shoppers were using just over 7 billion plastic bags a year. By the end of 2016 this number had dropped to 500 million a year. This is a significant reduction and more people are reusing their old plastic bags or using the stronger, more robust, 'bags for life'.
>
> Plastic bags are made using crude oil as the raw material. Crude oil is heated and then goes through further chemical reactions requiring energy and the combustion of chemicals.
>
> A leading scientist has claimed that the introduction of the 5p charge has made a huge positive impact on our use of resources and on the environment.

06.1 Evaluate the leading scientist's claim that the introduction of the 5p charge has had a positive impact.

Use both your own knowledge and the information in the magazine article to justify your opinion.

Continue your answer on the next page

[6 marks]

06.2 What is meant by the term **carbon footprint**?

[1 mark]

06.3 The Earth's early atmosphere contained a large amount of carbon dioxide.

Explain why the levels of carbon dioxide reduced significantly after that period.

[2 marks]

07

07.1 Ibuprofen is a chemical compound that has been specifically made to act as an anti-inflammatory medicine

What name is given to any substance with two or more ingredients mixed together in the correct quantities for a specific use?

[1 mark]

07.2 Ibuprofen has this structure:

What is the molecular formula of ibuprofen?

[1 mark]

07.3 Calculate how many moles of ibuprofen are present in 2.56 g of the compound.

Give your answer to 2 significant figures.

Moles of ibuprofen =

[3 marks]

07.4 A student suggested that if you added ethanoic acid to ibuprofen, one of the products would be carbon dioxide.

How would the student test to see if any of the products were carbon dioxide?

Describe the result of the test.

[2 marks]

08 Ammonia (NH₃) can be made from reacting hydrogen (H₂) with nitrogen (N₂).

The reaction is an equilibrium reaction.

08.1 Write a balanced symbol equation for the reaction outlined above.

[3 marks]

08.2 The formation of butanol from butene and water is also an equilibrium reaction.

$$C_4H_8(g) + H_2O(g) \rightleftharpoons C_4H_9OH(g)$$

Producing butanol is an exothermic reaction.

How would increasing the **pressure** in the reaction change the yield of butanol at equilibrium?

Give a reason for your answer.

[2 marks]

08.3 How would increasing the **temperature** in the reaction change the yield of butanol at equilibrium?

Give a reason for your answer.

[2 marks]

08.4 Why does the addition of a catalyst help companies speed up the rate of reaction for their products?

[1 mark]

END OF QUESTIONS

BLANK PAGE

BLANK PAGE

The Periodic Table

*The lanthanides (atomic numbers 58–71) and the actinides (atomic numbers 90–103) have been omitted.
The relative atomic masses of copper and chlorine have not been rounded to the nearest whole number.

AQA

GCSE

Combined Science: Trilogy H

SET B – Physics: Paper 5 Higher Tier

Author: Lynn Pharoah

Materials

Time allowed: 1 hour 15 minutes

For this paper you must have:
- a ruler
- a calculator
- the Physics Equation Sheet (found at the end of the paper).

Instructions

- Answer **all** questions in the spaces provided.
- Do all rough work in this book. Cross through any work you do not want to be marked.

Information

- There are 70 marks available on this paper.
- The marks for questions are shown in brackets.
- You are expected to use a calculator where appropriate.
- You are reminded of the need for good English and clear presentation in your answers.
- When answering questions 04.1 and 07.1 you need to make sure that your answer:
  – is clear, logical, sensibly structured
  – fully meets the requirements of the question
  – shows that each separate point or step supports the overall answer.

Advice

- In all calculations, show clearly how you work out your answer.

Name: _____

01

01.1 Which **two** statements correctly describe a substance melting?

Tick **two** boxes.

A physical change taking place ☐

A change which cannot be reversed taking place ☐

A chemical change taking place ☐

A reversible change taking place ☐

[2 marks]

01.2 Define what is meant by **specific latent heat**.

....

....

....

[2 marks]

01.3 **Figure 1.3** shows the heating graph for a substance that is heated at a constant rate.

Figure 1.3

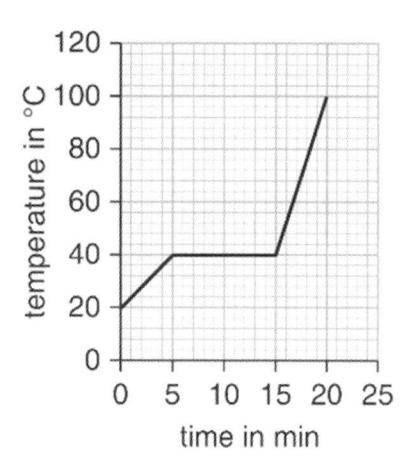

Determine the melting point of the substance from the graph.

Melting point = _____ °C

[1 mark]

01.4 The mass of the substance of question 01.3 is 0.090 kg

18 000 J of thermal energy is required to melt the substance.

Calculate the specific latent heat of fusion of the substance.

Use the correct equation from the Physics Equation Sheet.

_____

Specific latent heat of fusion = _____ J/kg

[3 marks]

02

02.1 Which **one** of the following components is an **ohmic conductor**?

Tick **one** box.

Filament bulb ☐

Fixed resistor ☐

Diode ☐

[1 mark]

02.2 A student wants to find out if a fixed length of wire behaves as an ohmic conductor.

She sets up a circuit to measure the current through the wire for various values of potential difference.

Crocodile clips at X and Y are used to connect to the ends of the wire.

Complete the circuit in **Figure 2.1** to enable her to take the measurements needed.

Figure 2.1

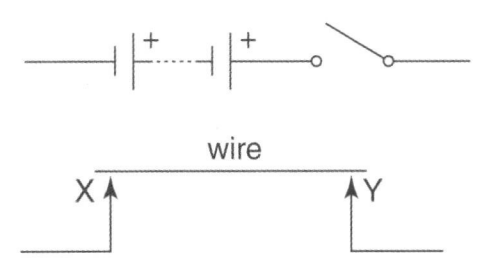

[3 marks]

02.3 **Figure 2.2** is a sketch graph of the student's current and potential difference data for the wire.

Figure 2.2

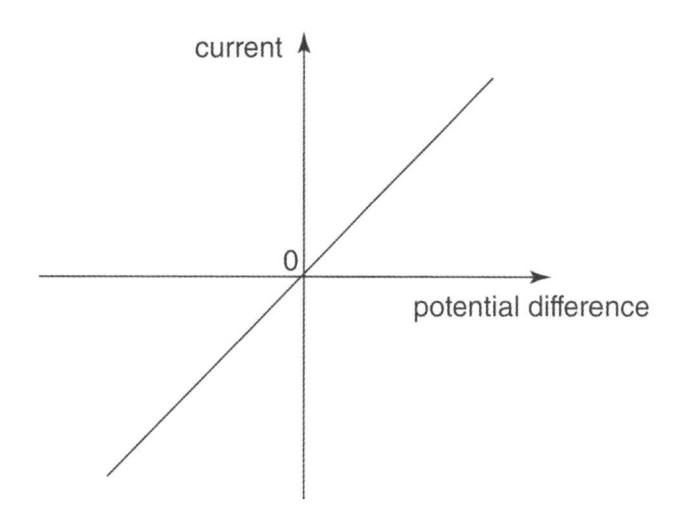

What can be concluded from **Figure 2.2** about the relationship between current and potential difference?

Is the wire an ohmic conductor?

....

....

....

[2 marks]

Question 2 continues on the next page

02.4 The student replaced the wire between the crocodile clips with a filament bulb.

Sketch the graph of current against potential difference that the student would expect for the filament bulb.

Use the axes in **Figure 2.3**.

Figure 2.3

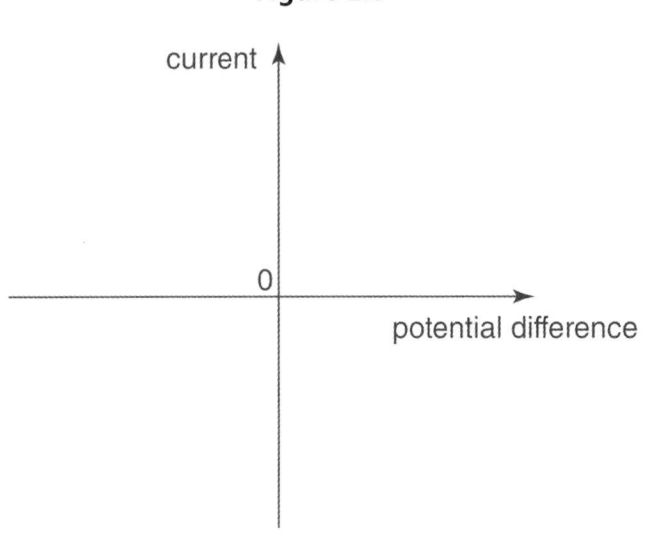

[2 marks]

02.5 Explain why the resistance of a filament bulb changes as the current through it is increased.

[2 marks]

03 A student is experimenting with a child's loop-the-loop track for toy cars, as shown in **Figure 3.1**.

Figure 3.1

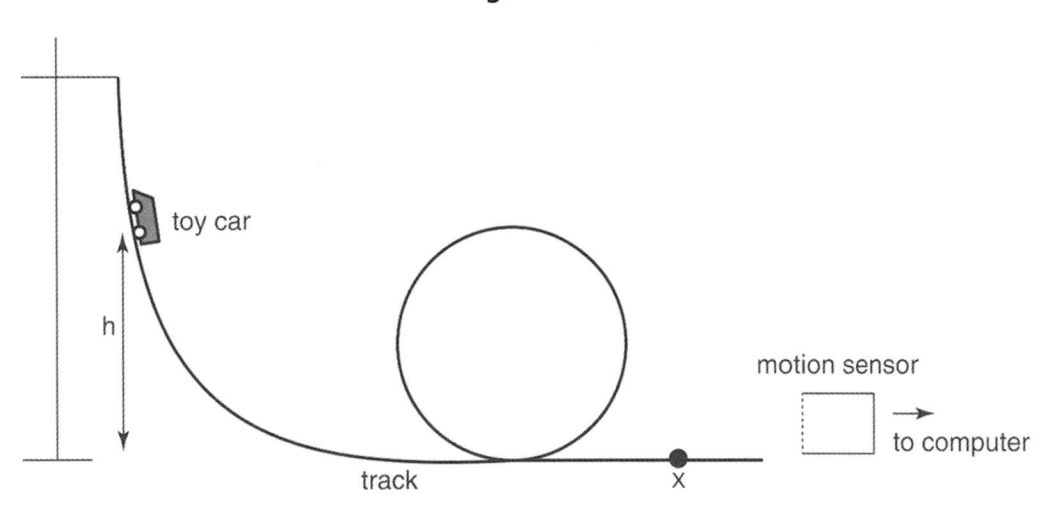

The student gradually increases the height (h) from which the toy car is released.

Eventually, the car has enough energy to complete the loop and travel along the track towards the motion sensor.

The motion sensor and computer record the speed of the toy car as it passes point X.

03.1 Write down the equation for gravitational potential energy.

[1 mark]

Question 3 continues on the next page

03.2 If the toy car is released at a height of h = 0.42 cm, it stays on the track and reaches the motion sensor.

The mass of the toy car is 50 g

Take gravitational field strength = 10 N/kg

Calculate the car's gravitational potential energy at height h = 0.42 cm

Gravitational potential energy = ... J

[3 marks]

03.3 Write down the equation for kinetic energy.

[1 mark]

03.4 At point X (**Figure 3.1**), the motion sensor records the car's speed as 2.0 m/s

Calculate the car's kinetic energy.

Kinetic energy = ... J

[2 marks]

03.5 Explain why the car's kinetic energy at X must be **less** than the car's gravitational potential energy at height h = 0.42 cm

[1 mark]

04

04.1 Compare the motion and arrangement of molecules when a substance is in its **solid** state, **liquid** state and **gas** state.

[4 marks]

04.2 State what is meant by the **internal energy** of a material.

[2 marks]

04.3 Describe **two** physical changes that would cause an **increase** in the internal energy of a fixed mass of water.

1.

2.

[2 marks]

Question 4 continues on the next page

04.4 **Figure 4.1** shows a sealed copper vessel containing air.

A gauge attached to the copper vessel shows the pressure of the air inside.

Figure 4.1

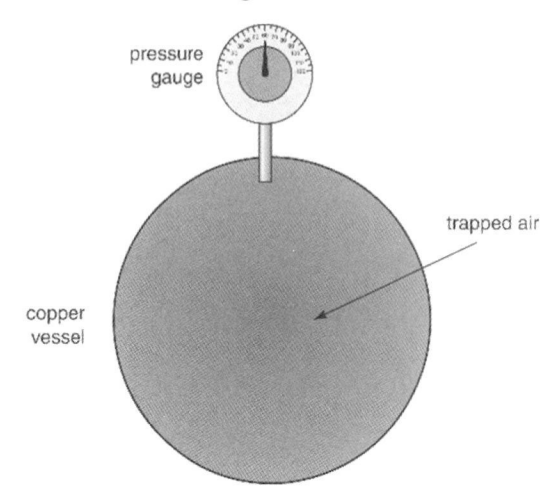

Describe how the air inside the copper vessel exerts pressure on the inside wall of the vessel.

[2 marks]

04.5 The copper vessel in **Figure 4.1** is lowered into a hot water bath.

This raises the temperature and the pressure of the air inside.

The volume of the copper vessel does not change.

Explain how the increase in temperature affects the pressure of the air inside the vessel.

[2 marks]

05

05.1 Explain what is meant by the **half-life** of a radioactive isotope.

[1 mark]

05.2 Radon is a radioactive gas.

Figure 5.1 shows the change in activity of a sample of radon-222 as time passes.

Figure 5.1

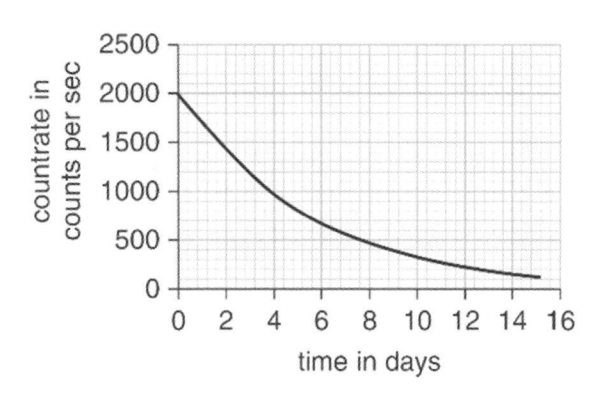

Use **Figure 5.1** to determine the half-life of radon-222.

Show your working.

Half-life = ... days

[3 marks]

Question 5 continues on the next page

05.3 The activity of a sample of radon gas is measured at 400 kBq

Calculate how many days would have to pass for the activity of the sample to be less than 50 kBq

Number of days = ...

[2 marks]

05.4 The naturally occurring isotope of radon is $^{222}_{86}$Rn, which decays by alpha emission to form polonium, Po.

Complete the following decay equation.

$$^{222}_{86}\text{Rn} \rightarrow {}^{218}_{\square}\text{Po} + {}^{\square}_{\square}\text{He}$$

[3 marks]

06 A temperature sensor inside an electric kettle measures the water temperature every 5.0 s

Figure 6.1 shows the graph of the temperature data that was collected.

Figure 6.1

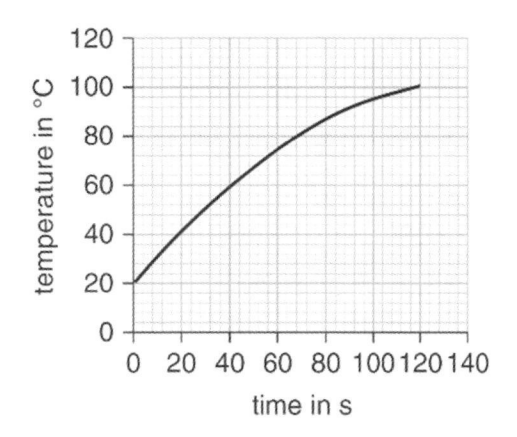

06.1 The gradient of the line in **Figure 6.1** represents the water's **rate of increase of temperature**.

Determine the water's rate of increase of temperature at 60 s

Show on the graph how you obtained your answer.

Give a suitable unit for your answer.

....

....

Rate of change of temperature = ...

Unit: ....

[3 marks]

Question 6 continues on the next page

06.2 What conclusion can be drawn from **Figure 6.1** about how quickly the water temperature rises as the water gets hotter?

....

....

[1 mark]

06.3 What does this suggest about the rate of dissipation of thermal energy to the surroundings as the water gets hotter?

....

....

[1 mark]

06.4 Table 6.1 shows the thermal conductivity of materials used for the outer casing of different kettles.

Table 6.1

Material	Thermal conductivity in W/(m K)
steel	16
polyethylene	0.33
polypropylene	0.22

Which material would be most effective at reducing the amount of thermal energy transferred from the kettle to the surroundings?

....

[1 mark]

07.1 Figure 7.1 shows the apparatus that can be used to measure the volume of a pebble, in order to determine the pebble's density.

The measuring cylinder measures up to 50 cm³ and has 1 cm³ graduations.

Figure 7.1

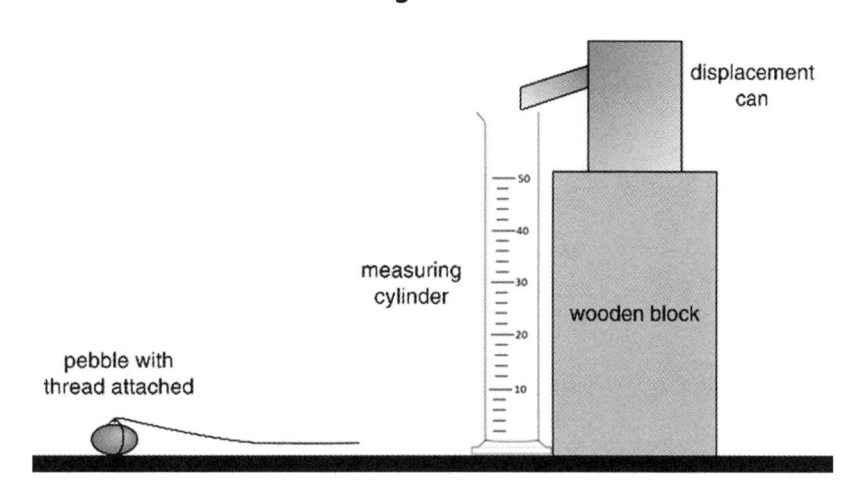

Write a series of instructions to determine the volume **and** the density of the pebble.

State any **additional** laboratory apparatus that may be required.

Suggest how errors can be kept as small as possible.

[6 marks]

Question 7 continues on the next page

07.2 A student suggests an alternative method to determine the volume of the same pebble.

He suggests lowering the pebble into a much larger measuring cylinder which already contains water **(Figure 7.2)**.

The resulting change in the water level would give the volume of the pebble.

Figure 7.2

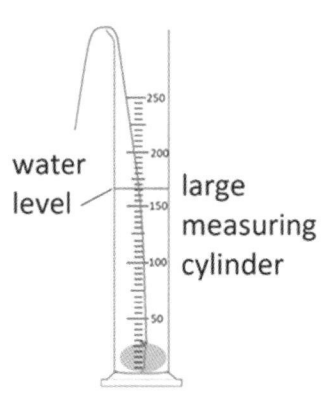

The student chooses a measuring cylinder that is large enough for the pebble to fit inside and measures up to 250 cm³ with 5 cm³ graduations.

Explain how the accuracy of the volume measurement using this method compares with the accuracy of the volume measurement using the method in **Figure 7.1**.

[2 marks]

08 A student uses the apparatus and the circuit shown in **Figure 8.1** to measure the specific heat capacity of a metal block.

Figure 8.1

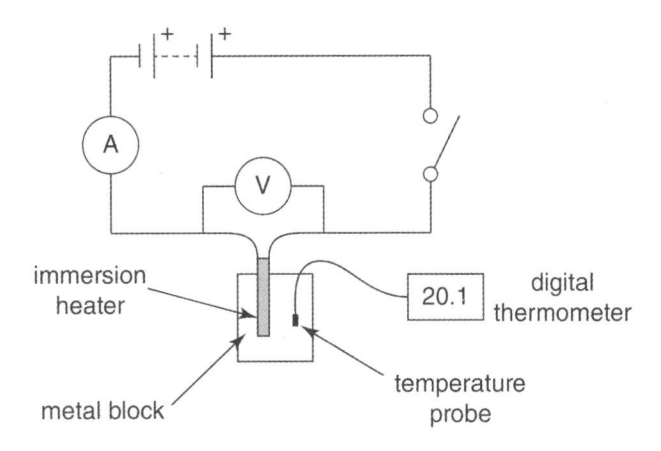

08.1 The battery supplies approximately 9 V of potential difference.

The resistance of the immersion heater is approximately 15 Ω

Calculate the approximate current drawn from the battery by the immersion heater when the switch is closed.

Current ≈ _____ A

[2 marks]

Question 8 continues on the next page

08.2 The student has three different ammeters, X, Y and Z, to choose from.

Details of these ammeters are shown in **Table 8.1**.

Table 8.1

Ammeter	Maximum current reading in A	Value of the smallest division in A
X	2	0.1
Y	1	0.05
Z	0.5	0.02

Which would be the most suitable ammeter to measure the current in the circuit?

Explain your answer.

Ammeter: ....

Explanation: ....

....

[3 marks]

08.3 The student closes the switch in the circuit of **Figure 8.1**.

The heater raises the temperature of the metal block.

The student's measurements are shown in **Table 8.2**.

Table 8.2

Quantity	Measurement
Mass of block	0.50 kg
Initial temperature	20.1 °C
Final temperature	30.1 °C
Current	0.55 A
Potential difference	8.0 V
Heating time	500 s

Calculate the specific heat capacity of the metal block.

Use the data in **Table 8.2**.

Specific heat capacity = ... J/kg°C

[4 marks]

END OF QUESTIONS

Physics Equation Sheet

Equation Number	Word Equation	Symbol Equation
1	(final velocity)² − (initial velocity)² = 2 × acceleration × distance	$v^2 - u^2 = 2\,a\,s$
2	elastic potential energy = 0.5 × spring constant × (extension)²	$E_e = \frac{1}{2} k e^2$
3	change in thermal energy = mass × specific heat capacity × temperature change	$\Delta E = m\,c\,\Delta\theta$
4	period = $\frac{1}{\text{frequency}}$	
5	force on a conductor (at right angles to a magnetic field) = magnetic flux density × current × length	$F = B\,I\,l$
6	thermal energy for a change of state = mass × specific latent heat	$E = m\,L$
7	potential difference across primary coil × current in primary coil = potential difference across secondary coil × current in secondary coil	$V_p I_p = V_s I_s$

AQA
GCSE
Combined Science: Trilogy H
SET B – Physics: Paper 6 Higher Tier
Author: Lynn Pharoah

Time allowed: 1 hour 15 minutes

Materials

For this paper you must have:
- a ruler
- a calculator
- the Physics Equation Sheet (found at the end of the paper).

Instructions

- Answer **all** questions in the spaces provided.
- Do all rough work in this book. Cross through any work you do not want to be marked.

Information

- There are 70 marks available on this paper.
- The marks for questions are shown in brackets.
- You are expected to use a calculator where appropriate.
- You are reminded of the need for good English and clear presentation in your answers.
- When answering questions 01.4 and 07.2 you need to make sure that your answer:
  – is clear, logical, sensibly structured
  – fully meets the requirements of the question
  – shows that each separate point or step supports the overall answer.

Advice

- In all calculations, show clearly how you work out your answer.

Name: _____

01 Infrared radiation is an electromagnetic wave.

01.1 Give the name of **two** types of electromagnetic wave that have a **shorter wavelength** than infrared radiation.

....

[2 marks]

01.2 Give the name of **one** type of electromagnetic wave that has a **longer wavelength** than infrared radiation.

....

[1 mark]

01.3 **Figure 1.1** shows the temperature of a surface and the amount of infrared energy radiated per second (the power) which radiates from it.

Figure 1.1

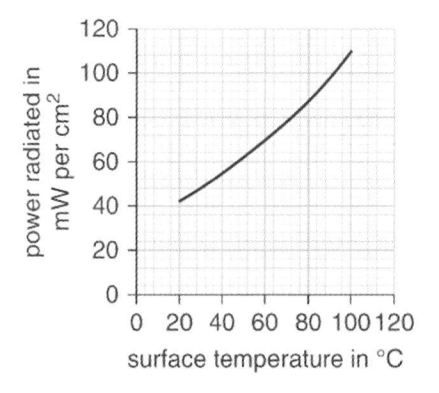

Give **two** conclusions based on the data shown in **Figure 1.1**.

1. ....

2. ....

[2 marks]

01.4 A student uses the apparatus in **Figure 1.2** to investigate the rate of emission of infrared radiation from different surfaces.

Figure 1.2

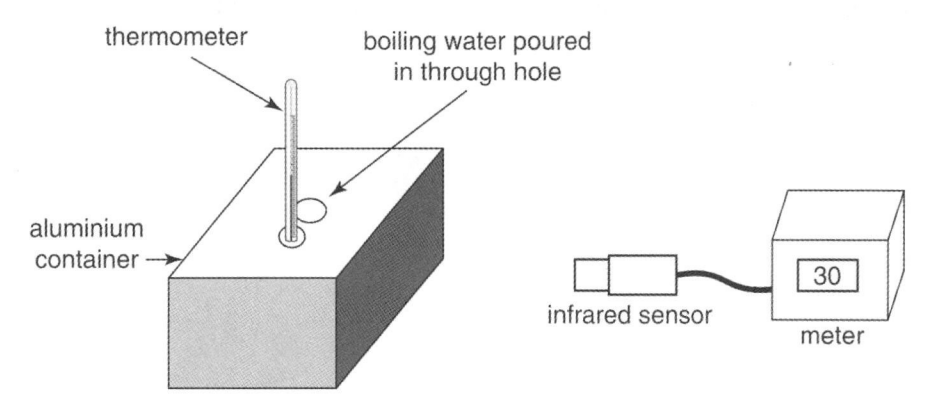

Boiling water poured into the aluminium container raises its temperature.

The infrared radiation from the container's surface is measured by the infrared sensor and displayed on the meter.

The four vertical faces of the container each have a different type of surface:

- shiny white
- dull white
- shiny black
- dull black.

Write a set of instructions for the student so that the amount of infrared radiated from each of the four surfaces can be compared fairly.

Identify any variables that must be controlled.

[4 marks]

Question 1 continues on the next page

01.5 **Table 1.1** shows data generated by the experiment.

Table 1.1

Surface	Shiny white	Dull white	Shiny black	Dull black
Energy radiated in mW/cm^2	3	23	30	33

Give **two** conclusions from the data.

1. _____

_____

2. _____

_____

[2 marks]

02

02.1 Describe the difference between a **vector** and a **scalar** quantity.

....

....

[1 mark]

02.2 Which **two** quantities listed are vectors?

Tick **two** boxes.

Mass ☐

Velocity ☐

Energy ☐

Momentum ☐

Work ☐

[2 marks]

Question 2 continues on the next page

02.3 The distance travelled by a cyclist during the first 5.0 seconds of a road race is monitored electronically.

Figure 2.1 shows the cyclist's distance–time graph for this first 5.0 s

Figure 2.1

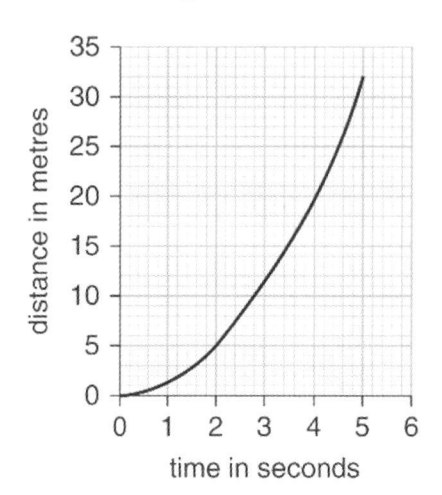

What can be concluded from **Figure 2.1** about the cyclist's motion during the first 5.0 s of the race?

Give a reason for your answer.

_____

_____

[2 marks]

02.4 Write down the equation that links distance travelled, speed and time.

_____

[1 mark]

02.5 Calculate the cyclist's **average speed** during the first 5.0 s of the race.

Average speed = ... m/s

[2 marks]

02.6 Use **Figure 2.1** to determine the cyclist's speed 3.0 s after the start of the race.

Show any working on **Figure 2.1** and below.

Speed = ... m/s

[3 marks]

Turn over >

03

03.1 Complete the following sentence using a term from the box.

| north pole | south pole | piece of magnetic material |

The direction of a magnetic field at any point is given by the direction of the force that would act on a _____ placed at that point.

[1 mark]

03.2 The wire in **Figure 3.1** passes vertically through a hole in a piece of cardboard, which is held horizontally.

The wire is connected to a cell.

The direction of the current in the wire is vertically downwards.

Figure 3.1

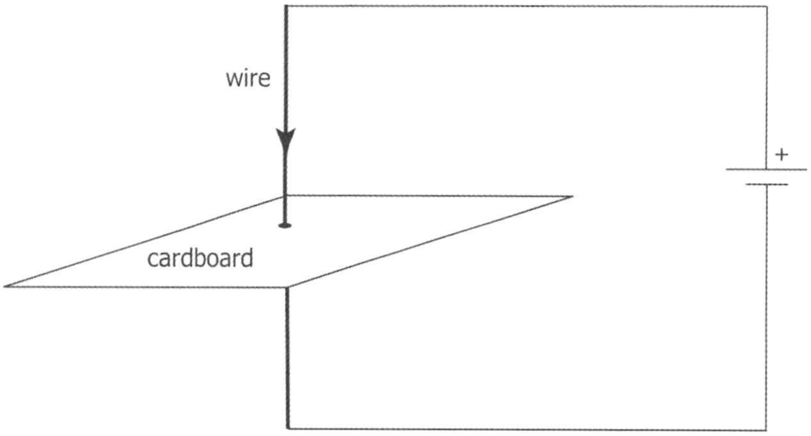

Describe how the shape of a magnetic field line produced around the wire in **Figure 3.1** can be plotted using a small magnetic plotting compass.

....

....

....

....

....

....

[3 marks]

03.3 **Figure 3.2** shows the cardboard **viewed from above**. The wire passes through the hole.

Figure 3.2

On **Figure 3.2**, draw the magnetic field produced by the wire.

You should draw **at least three** field lines.

Show the direction of the field.

[3 marks]

Question 3 continues on the next page

03.4 **Figure 3.3** shows the wire now wrapped around an iron core to create an electromagnet.

The wire is connected to a battery.

The electromagnet attracts an iron bar.

A hanger carrying a set of standard masses hangs vertically from the iron bar.

Figure 3.3

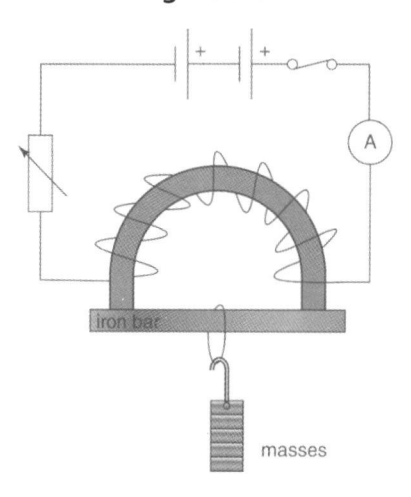

Masses are added to the hanger until the weight is too large for the electromagnet to support the iron bar. The iron bar and the masses then fall down.

A record of the masses is shown in **Table 3.1**.

Table 3.1

Object	Mass in g
Iron bar	250
Hanger	100
Additional mass before bar falls	650

Calculate the maximum attractive force exerted by the electromagnet.

Gravitational field strength = 9.8 N/kg

Maximum attractive force = _____ N

[2 marks]

03.5 A student discovers that:

- the maximum attractive force of the electromagnet is directly proportional to the **current** in the coil
- the maximum attractive force of the electromagnet is directly proportional to the **number of turns** on the coil.

The electromagnet has 20 turns on the coil and carries an electric current of 4.0 A

Predict the maximum attractive force that the electromagnetic could exert when there are **40 turns** on the coil and the current flowing is **8.0 A**

Maximum attractive force = ... N

[2 marks]

04

04.1 **Figure 4.1** is a snapshot of a wave on a long rope fixed to a wall at one end.

Figure 4.1

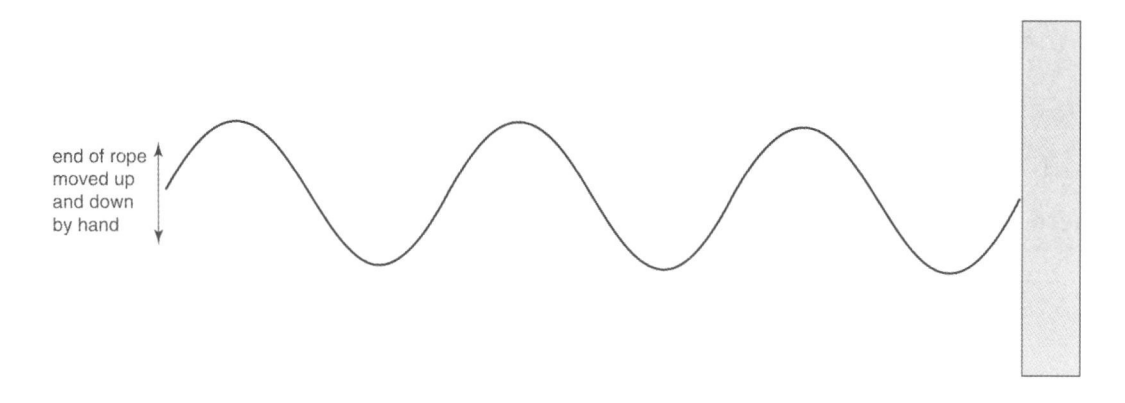

On **Figure 4.1**, show the distance corresponding to the **wavelength**.

Label this distance λ.

[1 mark]

04.2 On **Figure 4.1**, show the distance corresponding to the **amplitude**.

Label this distance **A**.

[1 mark]

04.3 Describe the difference between a **transverse** wave and a **longitudinal** wave.

[2 marks]

04.4 What type of wave is an electromagnetic wave?

Tick **one** box.

Transverse wave ☐

Longitudinal wave ☐

[1 mark]

04.5 The wavelength of some X-rays is 2.0×10^{-10} m

Electromagnetic waves travel at a speed of 3.0×10^{8} m/s in air.

Calculate the **frequency** of the X-rays.

Give your answer in standard form.

Frequency = ... Hz

[3 marks]

05

05.1 A van is moving forwards at a velocity of 5.0 m/s

The van has a mass of 1000 kg

Calculate the momentum of the van.

Give the unit for your answer.

Van's momentum = _____

Unit: _____

[3 marks]

05.2 The van collides into the back of a stationary car.

Explain how **Newton's Third Law** applies to this event.

[2 marks]

05.3 Write down what is meant by **conservation of momentum**.

[2 marks]

05.4 The collision between the van and the stationary car results in the van slowing down and the car moving forward.

Explain how the conservation of momentum applies to this collision.

[1 mark]

06 Newton's Second Law of motion states that "the acceleration of an object with constant mass is directly proportional to the resultant force acting on the object".

A student uses the air track and glider shown in **Figure 6.1** to demonstrate Newton's Second law.

Figure 6.1

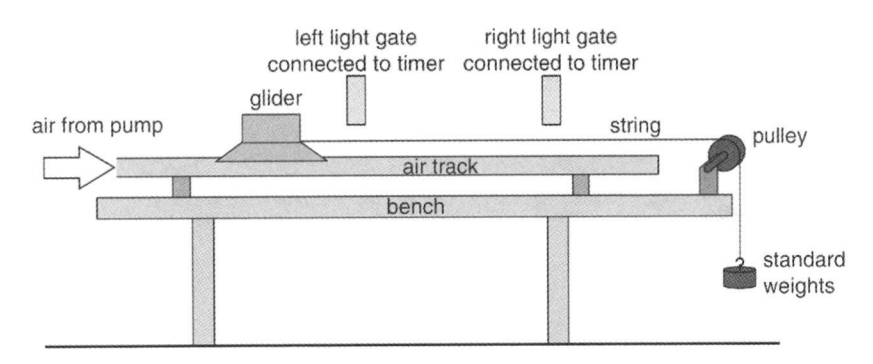

06.1 Air is pumped into the air track lifting the glider up from the track slightly.

Explain how this affects the motion of the glider along the track.

[2 marks]

Question 6 continues on the next page

06.2 The student attaches weights by string to the glider, to accelerate the glider along the track.

The times registered by the light gates (**Figure 6.1**) enable him to determine the acceleration of the glider.

He determines the acceleration for different weights attached to the string.

Identify the **independent**, the **dependent** and one **control variable**.

Independent variable: _____

Dependent variable: _____

Control variable: _____

[3 marks]

06.3 The student assumes that the total weight attached to the string is equal to the resultant force on the glider.

He uses the light gates in **Figure 6.1** to calculate the **initial velocity** and **the final velocity** of the glider.

One data set is shown in **Table 6.1**.

Table 6.1

Attached weight in N	Distance between light gates in m	Initial velocity in m/s	Final velocity in m/s
0.10	0.50	0.20	0.50

Calculate the acceleration of the glider.

Use the correct equation from the Physics Equation Sheet.

Give your answer to an appropriate number of significant figures.

Acceleration = _____ m/s^2

[3 marks]

06.4 The student takes more measurements to calculate the acceleration for different attached weights.

Figure 6.2 shows the resulting graph of acceleration versus attached weight.

Figure 6.2

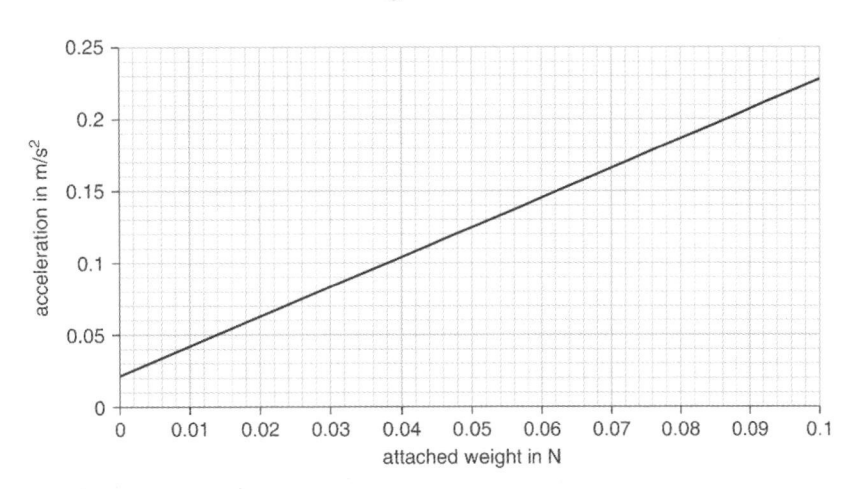

The student expected the results to show that the glider's acceleration is **directly proportional** to the attached weight.

Explain why the graph **does not** confirm what the student was expecting.

....

....

....

....

[2 marks]

06.5 Suggest **one** reason why the glider's acceleration was not directly proportional to the attached weight.

....

....

[1 mark]

Turn over >

07

07.1 The distance a car travels before it can be brought to a stop is known as **stopping distance**.

How is the stopping distance related to the driver's **thinking distance** and the car's **braking distance**?

[1 mark]

07.2 Explain what is meant by **thinking distance** and describe **three** factors, not including speed, that can affect thinking distance.

Explain what is meant by **braking distance** and describe **three** factors, not including speed, that can affect braking distance.

[6 marks]

07.3 **Figure 7.1** shows how thinking distance and braking distance depend on a car's speed.

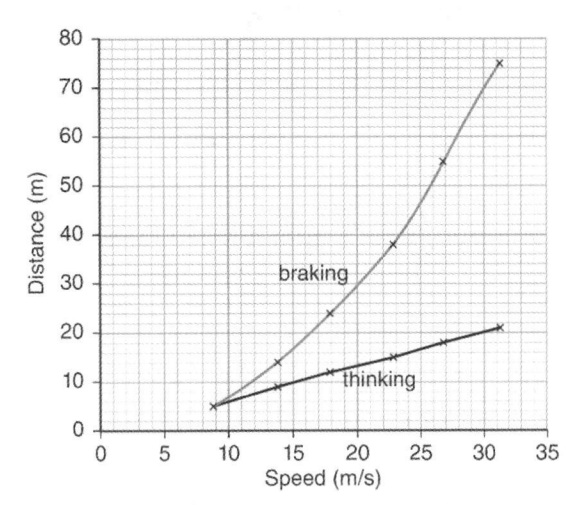

Figure 7.1

Compare the way that thinking distance and braking distance are affected by speed. Use information shown in **Figure 7.1**.

Include at least one similarity and at least one difference.

[3 marks]

END OF QUESTIONS

Physics Equation Sheet

Equation Number	Word Equation	Symbol Equation
1	(final velocity)² − (initial velocity)² = 2 × acceleration × distance	$v^2 - u^2 = 2as$
2	elastic potential energy = 0.5 × spring constant × (extension)²	$E_e = \frac{1}{2}ke^2$
3	change in thermal energy = mass × specific heat capacity × temperature change	$\Delta E = mc\Delta\theta$
4	period = $\frac{1}{\text{frequency}}$	
5	force on a conductor (at right angles to a magnetic field) = magnetic flux density × current × length	$F = BIl$
6	thermal energy for a change of state = mass × specific latent heat	$E = mL$
7	potential difference across primary coil × current in primary coil = potential difference across secondary coil × current in secondary coil	$V_p I_p = V_s I_s$

AQA

GCSE

Combined Science: Trilogy   H

SET B – Higher Tier

Authors: Kath Skillern, Paul Lewis and Lynn Pharoah

Answers

Acknowledgements

The author and publisher are grateful to the copyright holders for permission to use quoted materials and images.

All images are © HarperCollins*Publishers* and Shutterstock.com

Every effort has been made to trace copyright holders and obtain their permission for the use of copyright material. The author and publisher will gladly receive information enabling them to rectify any error or omission in subsequent editions. All facts are correct at time of going to press.

Published by Collins
An imprint of HarperCollins*Publishers*
1 London Bridge Street
London SE1 9GF

© HarperCollins*Publishers* Limited 2018
ISBN 9780008282769
First published 2018
10 9 8 7 6 5 4 3 2 1

All rights reserved. No part of this publication may be reproduced, stored in a retrieval system, or transmitted, in any form or by any means, electronic, mechanical, photocopying, recording or otherwise, without the prior permission of Collins.

British Library Cataloguing in Publication Data.

A CIP record of this book is available from the British Library.

Commissioning Editor: Rachael Harrison
Project Leader and Management: Natasha Paul and Katie Galloway
Authors: Kath Skillern, Paul Lewis and Lynn Pharoah
Cover Design: Paul Oates
Inside Concept Design: Ian Wrigley
Text Design and Layout: QBS Learning
Production: Lyndsey Rogers
Printed in Martins the Printers Ltd

Biology: Paper 1

Question	Answer(s)	Extra info	Mark(s)	AO/Spec ref.
01.1	tobacco mosaic virus	allow other viral disease, if correct	1	AO1 4.3.1.2
01.2	gives a distinctive 'mosaic' pattern of discolouration on the leaves which affects the growth of the plant due to lack of photosynthesis	allow other correct answers related to student's answer above	1 1 1	AO1 4.3.1.2
01.3	any two from: skin nose trachea/bronchi stomach phagocytic action		2	AO1 4.3.1.6
01.4	any two from: phagocytosis antibody production antitoxin production (allow descriptions instead)		2	AO1 4.3.1.6
02.1	enzymes		1	AO1 4.2.2.1
02.2	amylase, buffer, starch	must be correct order	1	AO2 4.2.2.1
02.3	buffer must be added to the enzyme before the starch is added – as the reaction will start as soon as the enzyme and starch meet if no buffer (or added afterwards) results will not be valid as the pH will be changed after the reaction has started		1 1	AO2 4.2.2.1
02.4	a control makes it easier to compare colours, as the water in the control doesn't contain any starch/so you can be sure all the starch is gone/digested/broken down, if it is the same colour as the control		1 1	AO2 4.2.2.1
03.1	to keep specimen flat to retain liquid under it	allow – to prevent the specimen touching the microscope lens	1 1	AO1 4.1.1.2

Question	Answer(s)	Extra info	Mark(s)	AO/Spec ref.
03.2	smaller field of view with a high-power lens **because** has greater magnification	or converse: larger with low-power lens because smaller magnification must state reason (i.e. because… for 2 marks, not just high is smaller and low is bigger)	1 1	AO1 4.1.1.2 4.1.1.5
03.3	scale bar – depends on size of student's diagram; the diameter at the point of the label of the meristem is about 400 mm; so, currently very approximately 10 mm = 200 μm	1 mark for drawing, with distinct meristem area 1 mark for label 1 mark for sensible units/scale (1 cm long scale. Units mm and μm) 1 mark for correct scale bar	2 2	AO2 4.2.3.1
04.1	accept values in range 65 000–70 000		1	AO2 4.2.2.7 4.2.2.5
04.2	active = nearly 4000 incidences (allow ± 1000) drink less alcohol = 12 000 incidences (allow ± 1000) and therefore drinking less alcohol produced about 3 times fewer cancers as being active	1 mark for both readings must include the comparison for second mark	1 1	AO2 4.2.2.7 4.2.2.5

Question	Answer(s)	Extra info	Mark(s)	AO/Spec ref.
04.3	men and women make different lifestyle choices		1	AO3 4.2.2.5 4.2.2.6 4.2.2.7
	men might not eat enough fruit and veg, are exposed to dangers at work and might drink too much alcohol	some indication of unhealthy lifestyle choice made by men	1	
	women might not exercise as much/are overweight, susceptible to human papilloma virus (HPV)	some indication of unhealthy lifestyle choice made by women	1	
05.1	lymphocytes instantly recognise live Lumpius/pathogen because it has the same antigens as the vaccine and respond **more quickly** to the infection by producing many specific antibodies which lock onto the Lumpius/pathogen and kill them before person becomes ill/person is immune/has immunity	accept white blood cells instead of lymphocytes must state 'more quickly' or equivalent must express concept that person does not become ill	3	AO2 4.3.1.6 4.3.1.7
05.2	efficacy – vaccine works/looks promising/passes to next stage of trial/positive result	must give reason for answers	1	AO3 4.3.1.9
	because many specific antibodies are produced when volunteers are infected with live Lumpius/pathogen	accept caution – insufficient data, adverse side effects/deaths – is dose too high? can acknowledge this thought process	1	
	dose – is good/correct		1	
	because response elicited (i.e. production of antibodies)		1	
05.3	clinical trial		1	AO3 4.3.1.9
	many volunteers recruited/tested on many humans		1	
06.1	oxygen	do not accept O$_2$	1	AO1 4.4.1.2

Question	Answer(s)	Extra info	Mark(s)	AO/Spec ref.
06.2	**Level 3:** a detailed and coherent explanation is provided with most of the relevant content, which demonstrates a comprehensive understanding of the investigation and the order in which it is carried out. The response gives logical steps, with reasons.		5–6	AO2 4.4.1.2
	Level 2: a detailed and coherent explanation is provided. The student has a broad understanding of the investigation. The response makes mainly logical steps with some reasoning.		3–4	
	Level 1: simple descriptions of the investigation are made along with reference to photosynthesis. The response demonstrates limited logical linking of points.			
	No relevant content		0	
	Indicative content • set up apparatus as in diagram • make sure plant is photosynthesising (can see bubbles of oxygen) • measure and record the temperature of water in beaker; the water is intended to maintain a constant temperature (buffer), so the temperature should be taken periodically and kept constant; controlling other variables • measure and place lamp a specified distance from apparatus – control of light intensity related to distance of lamp from apparatus (or use a variable brightness lamp) • carry out at several different distances of lamp (five distances) • allow plant to acclimatise to each new distance of the lamp/light intensity (2 mins) • record production rate of oxygen – count bubbles over given time period – 1 min/5 mins, at each distance • repeat three times for each distance of the lamp/light intensity • calculate mean production oxygen rate • light intensity not linearly related to distance			
06.3	lots of sunshine = lots of oxygen produced/high rate of photosynthesis	allow converse lack of sunshine/in shady area = lower rate of photosynthesis/less oxygen produced	1	AO3 4.4.1.2
	and therefore lots of oxygen = aids respiration for fish	allow converse in shade = not so good for fish	1	

©HarperCollins*Publishers* 2018   Combined Science Set B - Answers   3

Question	Answer(s)	Extra info	Mark(s)	AO/Spec ref.
06.4	inverse square law/inverse proportion		1	AO3 4.4.1.2
	as light intensity increases (distance between lamp and plant decreases) the volume of oxygen (or the rate of bubble production) increases	allow converse	1	
	this indicates that the rate of photosynthesis increases with light intensity	allow converse	1	
07.1	[graph showing O₂ consumption vs time with labels: O₂ deficit, O₂ requirement, resting O₂ consumption, steady-state O₂ consumption, EPOC, start exercise, end exercise, end recovery] recovery label can be indicated anywhere in the blue shaded EPOC area labels can be either on the graph shape, or correctly placed on the x and y axes.		5	AO2 4.4.2.2 4.4.2.1
07.2	blood flowing through the muscles transports the lactic acid to the **liver** where it is **converted** back **into glucose**		1 1	AO1 4.4.2.1 4.4.2.2
07.3	amount of **extra** oxygen the body needs (compared with resting) **after** exercise	must convey idea of **extra** oxygen	1	AO1 4.4.2.2 4.4.2.1
	to react with the accumulated lactic acid/ remove it from the cells		1	
08.1	substances are moved from a lower concentration solution to a more concentrated solution (against a concentration gradient)		2	AO1 4.4.2.1 4.1.3.3
08.2	**Level 3:** a detailed and coherent description is provided with most of the relevant content, which demonstrates a comprehensive understanding of metabolism and how living processes are linked. The response is logical.		5–6	AO1 4.4.2.3 4.4.1.3 4.4.2.1 4.2.2.1 4.4.1.1 4.1.3.1
	Level 2: a detailed and coherent description is provided. The student has a broad understanding of metabolism. The response makes mainly logical steps with some linkage.		3–4	

Question	Answer(s)	Extra info	Mark(s)	AO/Spec ref.
	Level 1: simple descriptions of living processes are made. The response demonstrates limited logical linking of points.		1–2	
	No relevant content		0	
	Indicative content <ul><li>conversion of glucose to starch, glycogen and cellulose</li><li>the formation of lipid molecules from a molecule of glycerol and three molecules of fatty acids</li><li>the use of glucose and nitrate ions to form amino acids which in turn are used to synthesise proteins</li><li>breakdown of excess proteins to form urea for excretion</li><li>uses of glucose produced in photosynthesis – respiration, storage, to produce fat or oil for storage, to strengthen the cell wall</li><li>used to produce amino acids for protein synthesis.</li></ul>			
09.1	wheat production greatly increased/ over doubled	must refer to both lines of graph	1	AO2 4.4.2.1
	but the area of land remained more or less the same		1	
09.2	approximately 25 years	do not accept 1988–1990	1	AO3 4.4.2.1
09.3	any one of: better control of factors limiting photosynthesis improved intensive farming methods/monocultures genetic modification of wheat		1	AO3 4.4.1.2

Biology: Paper 2

Question	Answer(s)	Extra info	Mark(s)	AO/Spec ref.
01.1	abiotic		1	AO 1 4.7.1.1
01.2	any two from: <ul><li>food</li><li>mates</li><li>territory</li><li>water</li></ul>	do not accept space	2	AO 1 4.7.1.1
01.3	interdependence		1	AO 1 4.7.1.1
01.4	a community in which all the species and environmental factors are in balance		1	AO 1 4.7.1.1
	so that population sizes remain fairly constant		1	

Question	Answer(s)	Extra info	Mark(s)	AO/Spec ref.
01.5	grass – primary producer mouse – primary consumer snake – secondary consumer hawk – tertiary consumer/top predator or grass – primary producer grasshopper – primary consumer lizard – secondary consumer hawk – tertiary consumer/top predator	role must be named to gain mark, not just name of organism	4	AO2 4.7.2.1
02.1	fugu and green spotted puffer		1	AO3 4.6.4
02.2	167.7 million years ago	must give units accept mya	1	AO3 4.6.4
02.3	insufficient evidence currently to be more accurate		1	AO3 4.6.3.2 4.6.4
02.4	either: fossils or DNA profiling or antibiotic resistance (in case of bacteria)		1	AO1 4.6.3.4 4.6.3.5
02.5	• individuals in a species show a wide range of **variation** due to differences in genes/ **mutations** occur randomly (i.e. not in response to the environment) • some variations / mutations are **advantageous** (most are not)/allow adverse point • individuals with characteristics most suited to the environment are **more likely to survive and reproduce**/ allow adverse point • the genes that allow these individuals to be successful are **passed to their offspring**/ allow adverse point		1 1 1 1	AO1 4.6.2.1 4.6.2.2
02.6	a recessive allele is only expressed if two copies are present/therefore no dominant allele present		1	AO1 4.6.1.4

Question	Answer(s)	Extra info	Mark(s)	AO/Spec ref.
03.1	sensible scales on correct axis correctly plotting points drawing line – joining points or line of best fit labels on axis - y axis – percentage of population who have Type 2 diabetes (%), and x axis – mean body mass (kg)		1 1 1 1	AO3 4.5.3.2
03.2	correlation/positive correlation, as mean body mass increases so does percentage/incidence of type 2 diabetes		1	AO3 4.5.3.2
03.3	**Level 3**: a detailed and coherent explanation is provided with most of the relevant content, which demonstrates a comprehensive understanding of the negative feedback system and how blood glucose concentrations differ in people with and without diabetes after a meal. The response gives logical steps, with reasons.		5–6	AO1 4.5.3.7 4.5.3.2
	Level 2: a detailed and coherent explanation is provided. The student has a broad understanding of the negative feedback system and diabetes. The response makes mainly logical steps with some reasoning.		3–4	
	Level 1: simple description of diabetes is made along with reference to the negative feedback system. The response demonstrates limited logical linking of points.		1–2	
	No relevant content		0	
	Indicative content glucose levels detected by pancreas and stimulated to release insulin in to blood also release of glucagon is suppressed insulin binds with receptors on cells cells take up glucose there are less of these receptors in the diabetic person glucose is converted into glycogen primarily in liver and muscle cells and so levels in blood are reduced after breakfast the concentrations of blood glucose increase, in both people but person with diabetes increases **much more**			

©HarperCollinsPublishers 2018   Combined Science Set B - Answers

Question	Answer(s)	Extra info	Mark(s)	AO/Spec ref.
	both their concentrations decrease during the morning, but person with diabetes decreases much more slowly			
04.1	any two from: • slower • act for long time • act more generally • chemical	accept adverse differences as long as they are credited to correct response type	2	AO1 4.5.2.1 4.5.3.1
04.2	**Level 3:** a detailed and coherent explanation is provided with most of the relevant content, which demonstrates a comprehensive understanding of the reflex arc and how the nervous system is made up of different neurones. The response gives logical steps, with reasons.		5–6	AO1 4.5.2
	Level 2: a detailed and coherent explanation is provided. The student has a broad understanding of the reflex arc and the nervous system. The response makes mainly logical steps with some reasoning.		3–4	
	Level 1: simple description of the nervous system is made along with reference to the reflex arc. The response demonstrates limited logical linking of points.		1–2	
	No relevant content		0	
	Indicative content • information from receptors passes along nerve cells (neurones) • as electrical impulses to the central nervous system (CNS) • reflex, so conscious part of brain not involved/coordinator is spinal cord; automatic and rapid • the CNS coordinates the response of effectors, which are muscles in the arm • stimulus – heat on hand; receptor heat receptors in skin; coordinator; effector – muscles contracting; response – quickly move hand away • reflex arc – sensory neurone carries signal to relay neurone and then motor neurone carries signal to effector • synapse – microscopic gap where two neurones meet chemicals released by a neurone and diffuse across the gap, so that the message continues into the next neurone, triggering an electrical signal			
04.3	84 × 4 = 336 336 – 85 – 87 – 83 = 81 ms	must state units for third mark	1 1 1	AO3 4.5.2.1

Question	Answer(s)	Extra info	Mark(s)	AO/Spec ref.
04.4	drinking alcohol slows down reaction times		1	AO3 4.5.2.1
05.1	any one from: • green plants • algae/weed • producers/primary producers		1	AO2 4.7.2.1
05.2	*T. sarasinorum* numbers increase and they eat lots of fish eggs		1	AO2 4.7.2.1 4.7.1.1 4.7.1.3
	therefore fewer fish survive from the eggs and there are fewer to eat, so 'elongated' eats more shrimp		1	
	'thicklip' numbers decrease as they are now in direct competition for shrimp, not enough shrimp for all		1	
05.3	live in different habitats (1 mark only)		1	AO2 4.7.2.1 4.7.1.1
	T. opudi lives in bush cover and rocks, whereas *T. wahjui* lives on the muddy bottom		1	
05.4	any one from: • sewage • fertilizer run-off • toxic chemicals • any named toxic chemical		1	AO1 4.7.3.2
06.1	**Level 3:** a detailed and coherent explanation is provided with most of the relevant content, which demonstrates a comprehensive understanding of protein synthesis and how it may be disrupted in Leigh syndrome. The response gives logical steps, with reasons.		5–6	AO2 4.6.1.5
	Level 2: a detailed and coherent explanation is provided. The student has a broad understanding of protein synthesis and that errors can cause the wrong protein to be made. The response makes mainly logical steps with some reasoning.		3–4	
	Level 1: simple descriptions of protein synthesis are made along with reference to errors. The response demonstrates limited logical linking of points.		1–2	
	No relevant content		0	
	Indicative content • proteins consist of chains of amino acids, coded for by a triplet of bases • each protein has a particular number and sequence of amino acids • if this is altered, then the wrong protein is made • transcription happens in the cell nucleus, where the DNA is copied • the two DNA strands unzip, complementary bases pair up with bases on the template strand • cytosine pairs with guanine, uracil pairs with adenine to form a strand of mRNA, which travels to the ribosome, where it is translated			

Question	Answer(s)	Extra info	Mark(s)	AO/Spec ref.
	• the ribosome reads off the triplet codes and carrier molecules bring specific amino acids to the protein chain in the correct order • the amino acids bond together to form a polypeptide chain, which folds to a specific shape to form a protein • Leigh syndrome could be a problem with unzipping, or a problem with transcription – the wrong base pairs with the template strand • or the ribosome may read the codon incorrectly or the carrier molecule may bring the wrong amino acid, all of which would cause the wrong protein to be made			
06.2	any one from: • search for genes linked to different types of disease • understanding and treatment of inherited disorders • use in tracing human migration patterns from the past	allow specific correct example	1	AO1 4.6.1.4
06.3	the gardener's method: • involves **selective breeding** • is the traditional method of breeding together individuals with desired characteristics • is the more natural method • takes a long time (many generations) • offspring won't definitely have trait the gardener wants the farmer's method: • involves **genetic engineering** • is more technical • is faster by transplanting specific genes for desired characteristics • is more expensive • offspring will definitely have the desired traits		2 (two points required) 2 (two points required)	AO2 4.6.2.3 4.6.2.4
07.1	population size means the number of individuals of a species that live in a habitat (number) population density is the number of individuals in a given/specific area		1 1	AO1 4.7.1.1
07.2	transect		1	AO2 4.7.1.1

Question	Answer(s)	Extra info	Mark(s)	AO/Spec ref.
07.3	systematic sampling: at regular intervals (e.g. every 50 cm) intervals must be sufficient to capture the changes in vegetative cover		1 1	AO2 4.7.1.1
07.4	construct further transects at 10 m intervals/ other sensible distance down the path take quadrats at the same distances as before (as suggested in 07.3) along these transects calculate the means at each quadrat place along the length of the path (add up all the plantains and divide by number of quadrats along the length of the path) to give mean number across the path		1 1 1	AO2 4.7.1.1
07.5	plants complete with each other for limited resources/many plants at verge, lots of competition plantain leaves are tough/have adapted to being trampled and may out-complete more delicate plants, which are trampled in the middle of the path		1 1	AO3 4.7.1 4.7.1.3 4.7.1.4

Chemistry: Paper 3

Question	Answer(s)	Extra info	Mark(s)	AO/Spec ref.
01.1	A reaction that transfers / releases energy to the surroundings and increases their temperature (warms/heats them up)	transfers / releases energy on its own is not enough and should not be awarded the mark	1	AO1 5.5.1.1
01.2	−5 Allow correct answer written in table instead.	one mark for 5, one mark for minus sign	2	AO2 5.5.1.2
01.3	temperature change		1	AO1 5.5.1.1
01.4	two marks for all four temperature changes drawn accurately (only accept temp changes, not start / end temps); only one mark for 3/4 temperature changes drawn accurately one mark for suitable x-axis including labelling one mark for suitable y-axis including labelling		4	AO2 5.5.1.2 WS3.1

Question	Answer(s)	Extra info	Mark(s)	AO/Spec ref.
01.5	data is not just continuous / numeric, or words to that effect		1	AO2 5.5.1.2 WS2.2
01.6	Investigation = C		1	AO3 5.5.1
	Explanation = any one from:		1	
	C only went up 10° so will get warm but not too hot.			
	D goes up by 28° which could burn the skin.			
	A does not increase enough.			
	B is an endothermic reaction.			
02.1	carbon		1	AO1 5.2.3
02.2	**Level 3** A linked explanation for both the electrical conductivity and high melting point. Spelling and grammar used with accuracy nearly all of the time.		5–6	AO2 5.2.3.1 5.2.3.2 5.2.3.3
	Level 2 Basic comment made regarding conductivity and melting point. Linked explanation for either electrical conductivity or melting point. Spelling and grammar used accurately most of the time.		3–4	
	Level 1 Basic comment about either conductivity or melting point. Spelling and grammar used with high levels of inaccuracy.		1–2	
	No relevant content		0	
	Indicative content			
	Electrical conductor			
	each carbon atom is bonded to three others			
	has a free / delocalised electron…			
	…which can carry a charge (through the structure)			
	High melting point / strong			
	giant structure			
	covalent bonds			
	strong bonds			
	require lots of energy to break them			
03.1	molten magnesium chloride	both answers required; more than two ticks negates the mark	1	AO1 5.4.3.1 5.4.3.2 5.4.3.4
	a solution of potassium iodide			
03.2	chlorine		1	AO2 5.4.3.4
03.3	hydrogen ions are positive and opposite charges attract	hydrogen ions are positive / opposite charges attract alone is not enough for the mark	1	AO2 5.4.3.1 5.4.3.1 5.4.3.4

Question	Answer(s)	Extra info	Mark(s)	AO/Spec ref.
03.4	sodium is more reactive than hydrogen	accept converse	1	AO1 5.4.3.4
	(sodium reacts with water to make sodium hydroxide; hydrogen does not react with water)			
03.5	hydrogen gains one electron		1	AO2 5.2.1.2
04.1	12 protons	ignore electrons when calculating mass number	1	AO2 5.1.1.5
	12 neutrons		1	
	mass number = protons + neutrons	accept mass number – atomic number = 12	1	
		accept mass number = number of particles in nucleus / all mass is in nucleus		
04.2	giant ionic		1	AO1 5.2.1.3
04.3	strong electrostatic forces of attraction (between ions)		1	AO2 5.2.1.3
	large amounts of energy needed to overcome / break the force		1	
04.4	M_r of MgO = 24 + 16 = 40	accept 14.75 ÷ 24 × 40	1	AO2 5.3.2.2
	40 ÷ 24 × 14.75		1	
	= 24.58		1	
04.5	14.755 (g)		1	AO3 5.3.1.4
04.6	to ensure that			AO2 5.3.2.4
	oxygen is not the limiting factor…		1	
	…so all of the magnesium reacts		1	
05.1	one pair of electrons shared		1	AO1 5.2.1.4
	three pairs of electrons remaining on outer shell of each atom			
05.2	same number of protons		1	AO2 5.1.1.5
	different number of neutrons		1	
05.3	(79 × 50) + (80 × 3) + (81 × 47) / 100		1	AO3 5.1.1.6
	= 79.97		1	
	= 80.0 to 3 s.f.		1	

Question	Answer(s)	Extra info	Mark(s)	AO/Spec ref.
05.4	**Level 3**: correct statement regarding trend in reactivity for both Group 1 and Group 7 with a linked explanation for both Group 1 and Group 7. Spelling and grammar used with accuracy nearly all of the time.		5–6	AO1 AO2 5.1.2.5 5.1.2.6
	Level 2: correct statement regarding trend in reactivity for both Group 1 and Group 7 with a linked explanation for Group 1 or Group 7. Spelling and grammar used accurately most of the time.		3–4	
	Level 1: simple correct statement regarding trend in reactivity or general statement about elements in either Group 1 or Group 7. Spelling and grammar used with high levels of inaccuracy.		1–2	
	No relevant content		0	
	Indicative content **General comments:** reactivity down Group 1 increases reactivity down Group 7 decreases group 1 want to lose electron group 7 want to gain electron further down either group atom gets bigger as there are more shells elements lower down each group have more shielding **Linked explanations:** further down Group 1 there are more shells so more shielding from nucleus electrons are more easily lost in atoms of elements lower down in Group 1 as there is less attractive force from nucleus to outer electron elements lower down in Group 1 have higher reactivity as they have more electron shells / energy levels further down Group 7 there are more shells so more shielding from nucleus electrons are more difficult to attract in atoms of elements lower down in Group 7 as there is less attractive force from nucleus to electron on neighbouring atom elements lower down in group 7 have lower reactivity as they have more electron shells / energy levels			
06.1	add excess magnesium to given volume of sulfuric acid		1	AO2 5.4.2.3
	filter off excess magnesium		1	
	(heat to) evaporate water / Leave to crystallise		1	
06.2	lit splint makes popping sound		1	AO1 5.8.2.1
06.3	24 + 32 + (16 × 4) = 120		1	AO2 5.3.1.2 5.3.2.1
	(32 ÷ 120) × 100 = 26.666 = 26.7% to 3 s.f.		1	

Question	Answer(s)	Extra info	Mark(s)	AO/Spec ref.
07.1	products at higher energy than reactants		1	AO1 5.5.1.2
	correct label for activation energy and overall energy change		1	
07.2	reactants = $2C_2H_6$ + $7O_2$ (2 × (348 + (6 × 413)) + (7 × 495) = 5652 + 3465 = 9117	allow ecf for final mark	1	AO2 5.5.1.3
	products = $6H_2O$ + $4CO_2$ (6 × (2 × 463)) + (4 × (2 × 799)) = 5556 + 6392 = 11 948		1	
	energy change = 9117 − 11 948 = −2831 kJ/mol		1	
07.3	more energy is released making / forming bonds…		1	AO2 5.5.1.1. 5.5.1.3
	…than energy is needed to break bonds.		1	
08.1	shells drawn with electron configuration of 2,8		1	AO1 5.1.1.7 5.2.1.2
	brackets and 3+ charge		1	
08.2	(flexible / malleable because) particles / ions / atoms are arranged in layers		1	AO2 5.2.2.7
	layers can slide over each other		1	
08.3	cryolite added to lower the melting point		1	AO2 5.4.3.3
	less energy needed to make aluminium oxide molten		1	
	at anode, oxygen reacts with carbon to form carbon dioxide		1	
	anode corrodes away		1	
08.4	$Al^{3+} + 3e^- \rightarrow Al$		1	AO1 5.4.3.5
08.5	reduction		1	AO2 5.4.3.5

Chemistry: Paper 4

Question	Answer(s)	Extra info	Mark(s)	AO/Spec ref.
01.1	suitable scale given on y-axis; all bars drawn correctly for 2 marks	y-axis should go up uniformly	1; 2	AO2 WS3.1
01.2	**independent variable:** area of UK water comes from; **dependent variable:** total volume of sodium hydroxide added; **control variable (any one from):** volume of water added; drops of universal indicator added; temperature of stated liquid; volume of sodium hydroxide added each time; concentration of sodium hydroxide used		1; 1; 1	AO3 WS3.2
01.3	to ensure all reactants are (completely) mixed	'particles' or 'chemicals' can be used instead of 'reactants'	1	AO3 WS2.2
02.1	carbon; hydrogen	or vice versa	1; 1	AO2 5.7.1.1
02.2	combustion	ignore 'complete' or 'incomplete'; accept 'exothermic'; do not accept 'burning'	1	AO1 5.9.3.1 5.5.1.1
02.3	$2C_8H_{18} + 25O_2 \rightarrow 16CO_2 + 18H_2O$	accept multiples or half these values (1, 12.5, 8, 9)	1	AO2 5.1.1.1
02.4	carbon monoxide; incomplete combustion / not enough oxygen in Reaction 1	accept 'not complete combustion'	1; 1	AO2 5.9.3.1
02.5	$C_{10}H_{20}$		1	AO1 5.7.1.4
02.6	high temperature; catalyst		1; 1	AO1 5.7.1.4
02.7	add bromine water; an alkene turns/stays colourless	reject 'clear'	1; 1	AO1 5.7.1.4

Question	Answer(s)	Extra info	Mark(s)	AO/Spec ref.
03.1	as temperature increases, time taken for magnesium to react decreases; the lower the temperature, the smaller the difference between two adjacent results	accept rate increases as temperature increases.; accept converse	1; 1	AO2 5.6.1.2
03.2	particles have more energy; particles move quicker/faster; frequency of successful collisions increases	ignore 'particles move more'; accept 'particles have more chance of collision'; ignore 'particles will collide more'	1; 1; 1	AO1 5.6.1.3
03.3	**Level 3:** workable plan including change of temperature and measurement of rate and fair testing and safety precautions.		5–6	AO2/3 5.6.1.2 5.6.1.3
	Level 2: a plan including change of temperature of acid and should include an attempt at measuring rate and / or an attempt at fair testing and safety.		3–4	
	Level 1: a simple plan without reference to changing any variable but should include an attempt at measuring rate or an attempt at fair testing.		1–2	
	No relevant content		0	
	Indicative content **Plan** • add magnesium to acid • time reaction / 'count bubbles' / measure volume of gas • valid measurements given • change temperature of acid • suitable safety measures in place **Control variables** • amount / mass / length / same 'size' of magnesium • volume / amount of acid			
04.1	any three from: screening/ grit removal to remove insoluble/ large substances; sedimentation to produce sewage sludge and effluent; anaerobic digestion of sewage sludge; aerobic biological treatment of effluent		3	AO1 5.10.1.3

10 Combined Science Set B - Answers

Question	Answer(s)	Extra info	Mark(s)	AO/Spec ref.
04.2	water that is safe to drink		1	AO1 5.10.1.2
04.3	any setup that involves a way of heating the sea water, condensing it, and collecting distilled/pure water	1 mark for sea water being heated by heat source 1 mark for labelled condenser 1 mark for distilled water	3	AO1 5.10.1.2
04.4	place damp blue litmus paper (into vapour)… …if litmus paper bleaches chlorine is present		1 1	AO1 5.8.2.4
05.1	any four from: crude oil is heated (most of the) oil evaporates vapours enter the column vapours cool **and** condense at their own/ unique boiling point top of column is cooler than bottom longer chains are removed from the column closer to the bottom more flammable substances leave at the top		4	AO1 5.7.1.2 5.7.1.3
05.2	the longer the carbon chain, the higher the boiling point	accept converse	1	AO2 5.7.1.3
05.3	(fractions are) mixtures / impure / not pure		1	AO2 5.8.1.1
05.4	water vapour — greenhouse effect sulfur dioxide — acid rain soot particles — global dimming	all three correct for 2 marks; two correct for 1 mark; no marks for only one correct	2	AO1 5.9.3.1 5.9.3.2
06.1	**Level 3:** a linked statement regarding a positive **and** negative impact of the 5p charge. Concludes with a balanced judgement (an evaluation) for full marks. Spelling and grammar used with accuracy nearly all of the time.		5–6	AO3 5.9.2.2 5.9.2.3 5.9.2.4 5.9.3.1 5.10.1.1 5.10.2.1 5.10.2.2
	Level 2: a linked statement regarding a positive **or** negative impact of the 5p charge. A simple statement regarding a positive **and** negative impact to the 5p charge. Spelling and grammar used with accuracy most of the time.		3–4	
	Level 1: basic comment about plastic bag use being in decline. Weak spelling and grammar.		1–2	
	No relevant content		0	
	Indicative content **Positive** • fewer plastic bags being used • less raw material/ crude oil being wasted • manufacturing plastic bags uses crude oil • crude oil is non-renewable • during production of plastic bags pollution will be given off • fewer plastic bags means lower carbon dioxide emissions, therefore less impact on the greenhouse effect and global warming • less global warming means reduced problems, e.g. flooding, animal habitats being destroyed • plastic bags are being reused • less use of landfill, which is good as plastic isn't biodegradable • less incineration, this is good as incineration causes toxic fumes which can be dangerous (to human health) **Negative** • bags being reused over time can become weaker and break • 'bags for life' still made from plastic, so crude oil is still being used • still 500 million bags being used in 2016 which required crude oil to make and pollution therefore occurred, as did using (finite) resources			
06.2	the total amount of carbon dioxide/ greenhouse gases given off by a substance/ product/ individual/ organisation during its life cycle		1	AO1 5.9.2.4
06.3	any two from: algae/ plants use carbon dioxide for photosynthesis carbon dioxide absorbed into oceans carbon dioxide involved in formation of sedimentary rocks	allow named example(s)	2	AO1 5.9.1.4
07.1	formulation		1	AO1 5.8.1.2
07.2	$C_{13}H_{18}O_2$ or $C_{12}H_{17}COOH$		1	AO2 5.2.1.4

Question	Answer(s)	Extra info	Mark(s)	AO/Spec ref.
07.3	M_r of $C_{13}H_{18}O_2$ = 206 2.56 ÷ 206 = 0.012 moles	allow ecf from incorrect M_r calculation; reject inverted calculation for marking point 2 reject final answer given with more/less than 2 significant figures	1 1 1	AO2 5.3.1.2 5.3.2.1
07.4	bubble gas through limewater… …limewater would go cloudy		1 1	AO1 5.8.2.3
08.1	$N_2 + 3H_2 \rightleftharpoons 2NH_3$	1 mark for correct reactants and products; 1 mark for correct balancing; 1 mark for use of reversible reaction arrow	3	AO2 5.6.2.1 5.6.2.3 5.1.1.1
08.2	increases the yield increasing pressure favours the side with fewer molecules, so equilibrium shifts to the right/ favours the forward reaction		1 1	AO1 5.6.2.7
08.3	decreases/ lowers the yield increasing temperature favours endothermic reaction, therefore favours reverse reaction		1 1	AO1 5.6.2.6
08.4	lowers the activation energy		1	AO1 5.6.1.4

Physics: Paper 5

Question	Answer(s)	Extra info	Mark(s)	AO/Spec ref.
01.1	a physical change taking place a reversible change taking place	1 mark each No more than 2 boxes ticked	2	AO1 6.3.1.2
01.2	(the amount of) energy required to change the state of 1 kg of a material with no change of temperature		1 1	AO1 6.3.2.3
01.3	40 (°C)		1	AO3 6.3.2.3

Question	Answer(s)	Extra info	Mark(s)	AO/Spec ref.
01.4	(energy = m L) 18000 = 0.090 × L $L = \dfrac{18000}{0.090}$ specific latent heat of fusion = 200 000 (J/kg)	1 mark for substitution into correct equation 1 mark for rearranging 1 mark for evaluation correct answer with no working 3 marks	3	AO2 6.3.2.3
02.1	fixed resistor	only one box ticked	1	AO1 6.2.1.4
02.2	ammeter in series with wire variable resistor in correct position to enable the current through the wire to be changed. voltmeter connected across the wire (allow ammeter and variable resistor to be in swapped places, or next to each other, as long as they are in series.)		1 1 1	AO1 6.2.1.4
02.3	current is directly proportional to potential difference yes (wire is an ohmic conductor)		1 1	AO3 6.2.1.4
02.4	curve through origin as shown negative section of line shown		1 1	AO1 6.2.1.4
02.5	as the current increases, the filament gets hotter filament resistance increases as its temperature increases		1 1	AO1 6.2.1.4
03.1	gravitational potential energy = mass × gravitational field strength × height	accept E = mgh	1	AO1 6.1.1.2

12 Combined Science Set B - Answers

Question	Answer(s)	Extra info	Mark(s)	AO/Spec ref.
03.2	50 g = 0.050 kg gravitational potential energy = 0.050 × 10 × 0.42 = 0.21 (J)	1 mark for unit conversion 1 mark for substitution 1 mark for evaluation correct answer with no working 3 marks	3	AO2 6.1.1.2
03.3	kinetic energy = 0.5 × mass × (speed)^2	accept $E = \frac{1}{2} m v^2$	1	AO1 6.1.1.2
03.4	kinetic energy = 0.5 × 0.05 × 2.0^2 0.10 (J) (accept 0.1)	1 mark for substitution 1 mark for evaluation correct answer with no working 2 marks	2	AO2 6.1.1.2
03.5	energy is dissipated/transferred to the surroundings//lost as thermal energy		1	AO1 6.1.2.1
04.1	**Level 2:** coherent account detailing the difference and similarities of solids, liquids and gases. Both molecule motion and arrangement included	3–4	4	AO1 6.3.1.1
	Level 1: relevant comments comparing solids and liquids, or liquids and gases both molecule motion and arrangement for maximum mark	1–2		
	No relevant content		0	
	Indicative content: molecules in solid and liquid much closer than in gases solid molecules vibrate solid molecules are joined together liquid and gas molecules move around, passing each other liquid and gas molecules move randomly gas particles move very rapidly			
04.2	total kinetic energy and potential energy (of all the atoms/molecules)	must include 'total' or 'sum of' kinetic and potential energy in either order	1 1	AO1 6.3.2.1

Question	Answer(s)	Extra info	Mark(s)	AO/Spec ref.
04.3	raise the temperature of the water convert the water to vapour		1 1	AO1 6.3.2.1
04.4	(air) molecules collide with the wall of the container exerting a force		1 1	AO1 6.3.3.1
04.5	air molecules move faster either: molecules exert a larger force (on the walls of the container or: molecules hit walls more frequently or: pressure increases	1 mark 1 mark for either statement	1 1	AO1 6.3.3.1
05.1	either: time taken for the number of nuclei to halve or: time taken for the activity/count rate to halve	1 mark for either statement	1	AO1 6.4.2.3
05.2	2000 to 1000: 3.8 days 1000 to 500: 7.6 – 3.8 = 3.8 days average = 3.8 days half-life = 3.8 (days)	1 mark for one set of data from graph or 2 marks for obtaining at least two sets of data and taking an average working must be shown 1 mark for correct half-life	3	AO3 6.4.2.3
05.3	3 half-lives would have to pass number of days = 3 × 3.8 = 11.4	1 mark 1 mark allow error carried forward from 05.2 for half-life value	2	AO2 6.4.2.3
05.4	$^{222}_{86}Rn \rightarrow ^{218}_{84}Po + ^{4}_{2}He$	1 mark for each correctly substituted number	3	AO2 6.4.2.2
06.1	tangent drawn on curve at time 60 s and large triangle used to calculate gradient rate of change of temperature = 0.75 (accept 0.65 to 0.85) unit: °C/s		1 1 1	AO2 6.1.2.1

©HarperCollins*Publishers* 2018 — Combined Science Set B - Answers 13

Question	Answer(s)	Extra info	Mark(s)	AO/Spec ref.
06.2	as the water gets hotter, the water's temperature rises less quickly		1	AO3 6.1.2.1
06.3	the rate of dissipation of thermal energy to the surroundings is increasing (as the water temperature gets hotter)		1	AO3 6.1.2.1
06.4	polypropylene		1	AO3 6.1.2.1
07.1	**Level 3:** a coherent plan covering all steps presented in a logical order detailing any additional apparatus used. The plan could be followed by another person to obtain a valid result for density. Additional equipment listed. Reference made to minimising errors.	5–6	6	AO2 6.3.1.1
	Level 2: a clear plan covering all major experiment steps presented in a logical order detailing any additional apparatus used. The plan could be followed by another person to obtain valid results for the mass and volume of the pebble. Additional equipment referred to.	3–4		
	Level 1: some relevant statements but the plan could not be followed by another person to obtain valid results.	1–2		
	No relevant content		0	
	Indicative content: beaker and additional measuring cylinder required fill the displacement can with water (and allow the excess to drain into a beaker) place the measuring cylinder under the spout of the displacement can lower the pebble into the can using the thread attached measure the volume of water in the measuring cylinder the volume of displaced water is equal to the volume of the pebble			

Question	Answer(s)	Extra info	Mark(s)	AO/Spec ref.
	repeat the process using a dry measuring cylinder of the same size calculate an average value for the volume of the pebble measure the mass of the (dry) pebble using a balance calculate the density of the pebble using $\rho = \dfrac{m}{v}$			
07.2	the size of each graduation on the larger measuring cylinder is bigger than on the smaller measuring cylinder the volume measurement using the larger measuring cylinder would be less accurate than the method using the displacement can		1 1	AO3 6.3.1.1
08.1	(V = IR) $9 = I \times 15$ $I = \dfrac{9}{15}$ current ≈ 0.6 (A)	1 mark for substitution into correct equation and rearranging 1 mark for evaluation correct answer with no working 2 marks	2	AO2 6.2.1.3
08.2	ammeter Y maximum current exceeds current drawn by immersion heater most accurate/best resolution in that current range	allow error carried forward from 08.1 in value of current drawn	1 1 1	AO3 6.2.1.3
08.3	energy transferred = 8.0 × 0.55 × 500 = 2200 J (E = mcΔθ) 2200 = 0.50 × c × (30.1 − 20.1) $c = \dfrac{2200}{0.50 \times (30.1 - 20.1)}$ specific heat capacity = 440 (J/kg°C)	1 mark for correct energy value 1 mark for substitution into correct equation from Equation Sheet 1 mark for rearranging 1 mark for evaluation correct answer with no working 4 marks	4	AO2 6.1.1.3

Physics: Paper 6

Question	Answer(s)	Extra info	Mark(s)	AO/Spec ref.
01.1	any two from: visible light, ultraviolet, X-rays, Gamma rays	any two for 1 mark each	2	AO1 6.6.2.1
01.2	any one from: microwaves, radio waves	any one for 1 mark	1	AO1 6.6.2.1
01.3	power radiated increases as the surface temperature increases		1	AO3 6.6.2
	power radiated increases more quickly/at a greater rate at higher temperatures		1	
01.4	**Level 2:** a coherent set of instructions presented in a logical sequence that would result in a set of data that would enable the infrared emission rates from the different surfaces to be compared fairly. The need for surface temperature as a control variable must have been considered for the maximum mark.	3–4	4	AO2 6.6.2.2
	Level 1: some relevant content but may not be clear how to make a fair comparison of emission rates from the data generated	1–2		
	No relevant content		0	
	Indicative content: position the infrared sensor at a specific distance from one of the container's surfaces			
	record an infrared sensor reading at a specific temperature (of the water in the container)			
	repeat the above steps with the infrared sensor pointing at each of the other three surfaces in turn checking the water temperature remains constant.			
	distance from sensor to container and the temperature at which the meter reading is taken must be the same for all four surfaces			

Question	Answer(s)	Extra info	Mark(s)	AO/Spec ref.
01.5	any two from: different surfaces emit different amounts of energy/infrared radiation (per second)	any two for 1 mark each	2	AO3 4.6.2.2
	a dull surface emits more energy/infrared radiation (per second) than a shiny surface of the same colour			
	black surfaces emit more energy/infrared radiation (per second) than white surfaces			
	the polished aluminium surface emits far more infrared radiation than any other surface tested			
2.1	a vector has magnitude/size and direction, a scalar has only magnitude/size		1	AO1 6.5.1.1
02.2	velocity	1 mark each	2	AO1 6.5.1
	momentum	no more than two boxes ticked		
02.3	increasing speed or accelerating		1	AO3 6.5.4.1.4
	because gradient is increasing		1	
02.4	distance (travelled) = speed × time		1	AO1 6.5.4.1.2
02.5	32 = average speed × 5.0	1 mark for substitution and rearranging	2	AO2 6.5.4.1.2
	average speed = $\frac{32}{5.0}$	1 mark for evaluation		
	average speed = 6.4 (m/s)	correct answer with no working 2 marks		
02.6	attempt to determine gradient at time 3.0 s by drawing a tangent to the curve and using a large triangle.	1 mark for correct additions to graph	3	AO2 6.5.4.1.4
	correct data selection from gradient triangle	1 mark for correct data selection		
	speed in the range 6.9 to 7.6 (m/s)	1 mark for correct evaluation of speed		
03.1	north pole		1	AO1 6.7.1.2

Question	Answer(s)	Extra info	Mark(s)	AO/Spec ref.
03.2	place the (plotting) compass on the cardboard and mark a dot (on the cardboard) next to the compass point		1	AO2 6.7.1.2
	reposition the compass over the dot and mark a dot next to the compass point		1	
	repeat the procedure and join the dots to form a complete magnetic field line		1	
03.3	at least three concentric circles drawn		1	AO2 6.7.2.1
	separation of circles increases with increasing distance from wire		1	
	arrows (on at least one wire) point clockwise		1	
03.4	total mass = 250 + 100 + 650 = 1000 g maximum attractive force (= weight) 1.0 × 9.8 = 9.8 N	1 mark for adding masses 1 mark for evaluation of max force correct answer with no working 2 marks	2	AO2 6.7.2.1 & 6.5.1.3
03.5	maximum attractive force = 9.8 × 2 × 2 = 39.2 (N)	1 mark 1 mark correct answer with no working 2 marks allow error carried forward from 03.3	2	AO2 6.7.2.1
04.1	wavelength marked correctly with arrows and labelled λ wavelength can be between any equivalent points on adjacent wave (see below for examples indicated by the arrows)		1	AO1 6.6.1.2
04.2	amplitude marked correctly with arrows and labelled A amplitude must be clearly between midpoint and peak or trough (examples shown below)		1	AO1 6.6.1.2

Question	Answer(s)	Extra info	Mark(s)	AO/Spec ref.
04.3	transverse wave: crest and troughs (or 'movement (of medium) is at right angles to direction of wave motion')		1	AO1 6.6.1.1
	longitudinal wave: compressions and rarefactions (or 'movement (of medium) is along the line of wave motion)		1	
04.4	transverse		1	AO1 6.6.2.1
04.5	(v = fλ) $3.0 \times 10^8 = f \times 2.0 \times 10^{-10}$ $f = \dfrac{3.0 \times 10^8}{2.0 \times 10^{-10}}$ frequency = 1.5 × 10^{18} (Hz)	1 mark for substitution into correct equation 1 mark for rearranging 1 mark for evaluation correct answer with no working 3 marks	3	AO2 6.6.1.2
05.1	(momentum = mv) Van's momentum = 1000 × 5.0 = 5000 unit: kg m/s	1 mark for substitution into correct equation 1 mark for evaluation correct answer with no working 2 marks 1 mark for unit	2 1	AO2 6.5.5.1 AO1 6.5.5.1
05.2	the size of the force exerted by the van on the car is equal to the size of the force exerted by the car on the van the force exerted by the van on the car is in the opposite direction to the force exerted by the car on the van	1 mark 1 mark (maximum 1 mark for simple statement of 'equal and opposite')	2	AO1 6.5.4.2.3
05.3	total momentum before (collision) = total momentum after in a closed system	1 mark additional 1 mark if this phrase is included	2	AO1 6.5.5.2
05.4	momentum of the van before the collision is equal to the total/sum of momentum of the van and the car after the collision		1	AO1 6.5.5.2

Question	Answer(s)	Extra info	Mark(s)	AO/Spec ref.
06.1	the glider moves freely/more smoothly (along the track)		1	AO1 6.5.4.2.2
	because friction is reduced/removed		1	
06.2	independent variable: resultant force (on glider)		1	AO3 6.5.4.2.2
	dependent variable: acceleration		1	
	control variable: mass (of glider) (accept: same glider or same air track set up)		1	
06.3	($v^2 - u^2 = 2as$) $0.50^2 - 0.20^2$ = 2 × acceleration × 0.50 acceleration = $\dfrac{0.50^2 - 0.20^2}{(2 \times 0.50)}$ acceleration = 0.21 (m/s^2)	1 mark for substitution into correct equation 1 mark for rearranging 1 mark for evaluation do not accept 0.2	3	AO2 6.5.4.1.5
06.4	(for the results to be as expected by the student) the graph should be a straight line passing through the origin		1	AO3 6.5.4.2.2
	line misses the origin / has a non-zero y-intercept		1	
06.5	any one from: incorrect to assume that the attached weight is equal to the resultant force. friction is having an effect the mass of the attached weight is also being accelerated so mass isn't constant	any one for 1 mark accept any other sensible suggestion.	1	AO3 6.5.4.2.2
07.1	stopping distance = thinking distance + braking distance (or this written as a sentence)		1	AO1 6.5.4.3.1
07.2	**Level 3:** a coherent account explaining both thinking and braking distance with the effect of at least three factors on each considered	5–6	6	AO1 6.5.4.3.2 6.5.4.3.3 6.5.4.3.4
	Level 2: a clear account explaining both thinking and braking distance with the effect of two factors on each considered.	3–4		
	Level 1: some relevant comments but lacks detail	1–2		

Question	Answer(s)	Extra info	Mark(s)	AO/Spec ref.
	No relevant content		0	
	Indicative content: thinking distance is the distance travelled by the car during the time that the driver is reacting to an emergency. braking distance is the distance travelled while the brakes are being applied. thinking distance can increase if: • the driver is tired • the driver has consumed drugs or alcohol • the driver is distracted by other people in the car or by events going on outside the car braking distance can increase if: • the road surface is wet or icy • the car's tyres are worn • the car's brakes are in poor condition • the gradient of the road			
07.3	both thinking distance and braking distance increase with speed		1	AO3 4.5.6.3
	thinking distance increases steadily with speed		1	
	braking distance increases at an increasing rate with increasing speed		1	

©HarperCollins*Publishers* 2018   Combined Science Set B - Answers

BLANK PAGE

AQA

GCSE

Combined Science: Trilogy H

SET A – Biology: Paper 1 Higher Tier

Author: Mike Smith

Time allowed: 1 hour 15 minutes

Materials

For this paper you must have:
- a ruler
- a calculator

Instructions

- Answer **all** questions in the spaces provided.
- Do all rough work in this book. Cross through any work you do not want to be marked.

Information

- There are 70 marks available on this paper.
- The marks for questions are shown in brackets.
- You are expected to use a calculator where appropriate.
- You are reminded of the need for good English and clear presentation in your answers.
- When answering questions 03.1, 06.2 and 07.2 you need to make sure that your answer:
  – is clear, logical, sensibly structured
  – fully meets the requirements of the question
  – shows that each separate point or step supports the overall answer.

Advice

- In all calculations, show clearly how you work out your answer.

Name: _____

01　The human body protects itself against pathogens in different ways.

01.1　Draw **one** line from each part of the body to the way it protects the body against pathogens.

Part of the body	Way it protects the body
Platelets	Acts as a barrier
Skin	Forms clots to seal wounds
Stomach	Secretes mucus to trap pathogens
Trachea and bronchi	Produces acid to kill pathogens

[3 marks]

01.2　Describe **one** advantage and **one** disadvantage of using antibiotics against pathogens.

Advantage: ....

....

Disadvantage: ....

....

[2 marks]

01.3 Explain how vaccination can protect the body against illness caused by pathogens.

[4 marks]

02 Most organisms consist of many cells.

02.1 Figure 2.1 shows an image of a white blood cell.

Figure 2.1

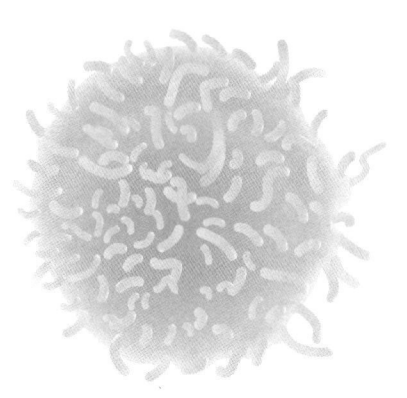

The diameter of the image is 60 mm

The image has been magnified 5000 times.

Calculate the actual size of the cell in μm

Use the formula:

$$\text{magnification} = \frac{\text{size of image}}{\text{size of real object}}$$

Actual size: _____ μm

[4 marks]

02.2 Electron microscopes can be used to view sub-cellular structures in detail.

Electron microscopes have a greater resolution (resolving power) than light microscopes.

Explain the difference between **resolution** and **magnification**.

[2 marks]

Question 2 continues on the next page

02.3 **Figure 2.2** shows a single-celled organism called *Euglena*.

Figure 2.2

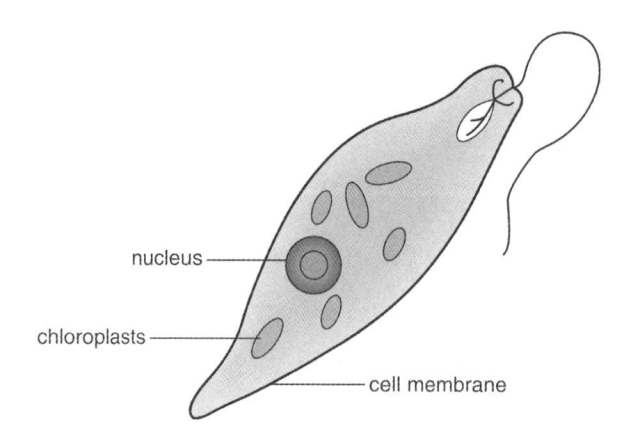

Euglena has been classified as a protist.

Suggest why it has **not** been classified as an animal, a plant or a bacterium.

It is **not** an animal because ....

....

It is **not** a plant because ....

....

It is **not** a bacterium because ....

....

[3 marks]

03 Amylase is an enzyme that digests starch to sugar.

The following method can be used to investigate the effect of pH on the activity of amylase.

1. Mix amylase solution and starch suspension in a boiling tube.
2. Put the boiling tube into a water bath at 25 °C
3. Remove a drop of the mixture every 30 seconds and test it for the presence of starch.
4. Repeat the investigation at different pH values.

03.1 The activity of amylase is also affected by temperature.

Use the method above to describe how you would investigate this. In your method, explain how you would identify the optimum temperature for amylase activity.

You should include:

- what you would measure
- variables you would control.

[6 marks]

Question 3 continues on the next page

03.2 Complete **Table 3.1**.

Table 3.1

Enzyme	Substrate	Product
Amylase	Starch	Sugar
Protease		
Lipase		

[4 marks]

03.3 Describe how to test for sugars.

....

....

....

[3 marks]

04 Rose black spot is a fungal disease which affects plants.

It causes purple or black spots on leaves.

The leaves then often turn yellow and drop early.

04.1 Plants infected with rose black spot grow much more slowly than plants that are **not** infected.

Explain why the infected plants grow more slowly.

[3 marks]

04.2 Give **two** methods to prevent rose black spot from spreading, without destroying the plants.

Explain how each method works.

Method 1:

Explanation:

Method 2:

Explanation:

[4 marks]

Turn over >

05 A student investigated osmosis using pieces of potato and sucrose solutions of different concentrations.

This is the method used:

1. Cut pieces of potato of the same size and shape.
2. Measure the mass of each piece.
3. Leave each piece of potato in a different concentration sucrose solution for one hour.
4. Remove each piece of potato, dry it with a cloth and measure its mass again.

Table 5.1 shows the student's results.

Table 5.1

Concentration of sucrose in mol per dm³	Mass of potato piece before being put in solution in g	Mass of potato piece after being put in solution in g	Percentage change in mass %
0.0	24.1	31.6	+31.1
0.2	24.0	29.0	
0.4	24.2	23.7	−2.1
0.6	23.9	19.3	−19.2
0.8	24.1	19.0	−21.2

05.1 Calculate the percentage change in mass for the potato in the 0.2 mol per dm³ sucrose solution.

Percentage change: _____ %

[3 marks]

05.2 Plot the data from **Table 5.1** and your answer to **Question 05.1** onto **Figure 5.1**.

Draw a smooth curved line of best fit.

[3 marks]

Figure 5.1

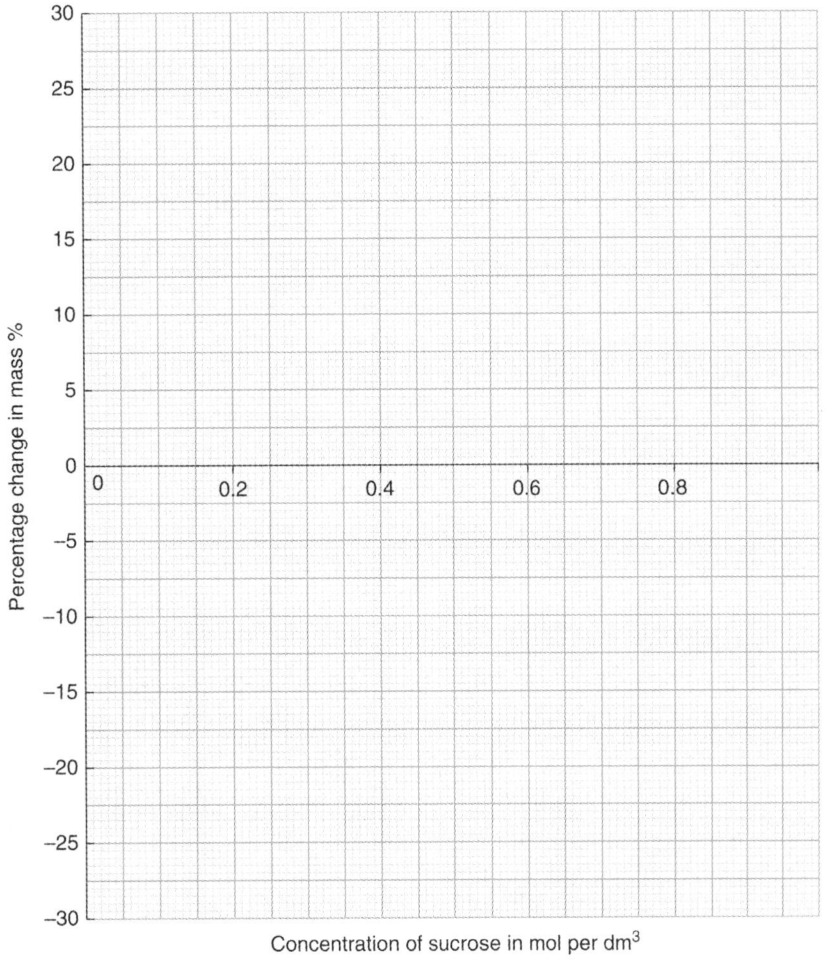
Percentage change in mass %
30
25
20
15
10
5
0
–5
–10
–15
–20
–25
–30
0
0.2
0.4
0.6
0.8
Concentration of sucrose in mol per dm3

05.3 Look at **Figure 5.1**.

What concentration of sucrose would have the same concentration of water molecules as in the pieces of potato?

Answer: .... mol per dm^3

[1 mark]

Question 5 continues on the next page

05.4 Why was it important to cut all the pieces of potato to be as near as possible the same size and shape?

[2 marks]

05.5 Why was it important to dry each piece of potato before measuring its mass a second time?

[1 mark]

06 Many diseases can be affected by lifestyle factors such as diet.

06.1 Explain why a diet that is too high in fat may lead to coronary heart disease.

[4 marks]

Question 6 continues on the next page

06.2 **Figure 6.1** shows data for obesity and Type 2 diabetes.

Figure 6.1

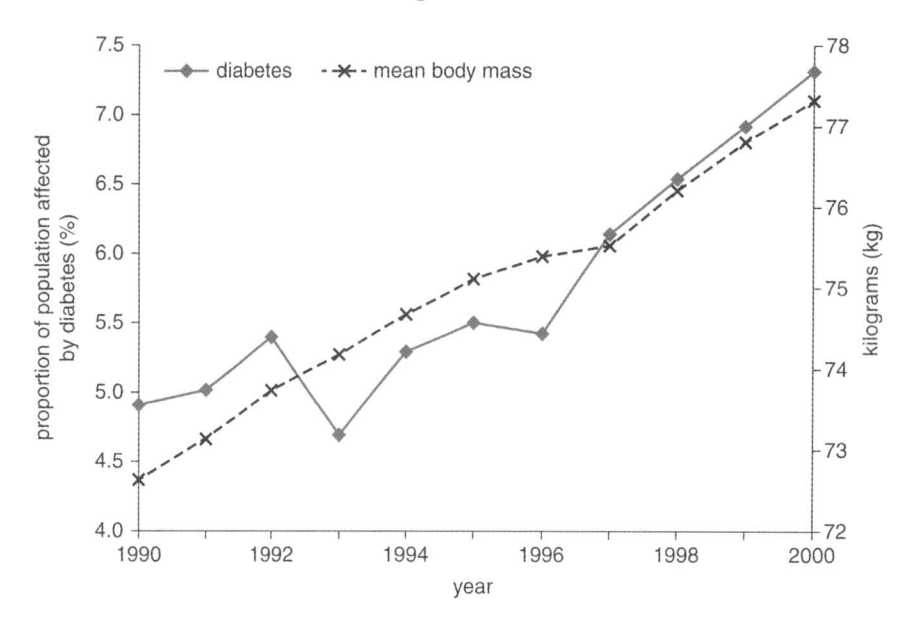

Evaluate whether the data from **Figure 6.1** shows that obesity is a risk factor for Type 2 diabetes.

[6 marks]

07 Figure 7.1 shows the effects of different limiting factors on the rate of photosynthesis.

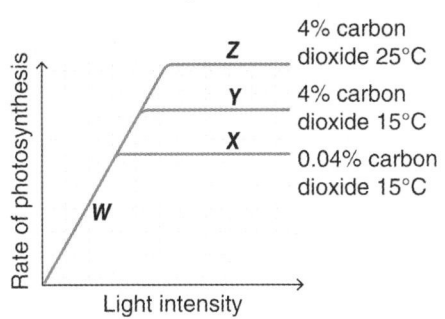

Figure 7.1

Graph: Rate of photosynthesis vs Light intensity, showing three curves labelled with points W (rising portion), X (plateau at 0.04% carbon dioxide 15°C), Y (plateau at 4% carbon dioxide 15°C), and Z (plateau at 4% carbon dioxide 25°C).

07.1 Identify the limiting factors at points **W**, **X** and **Y** on **Figure 7.1**.

Explain the reasons for your answers.

Limiting factor at point **W**: ....

Explanation: ....

Limiting factor at point **X**: ....

Explanation: ....

Limiting factor at point **Y**: ....

Explanation: ....

[6 marks]

Question 7 continues on the next page

07.2 Without more information, it is **not** possible to identify the limiting factor at point **Z**.

Explain how you could identify the limiting factor at point **Z**.

[6 marks]

END OF QUESTIONS

AQA
GCSE
Combined Science: Trilogy H
SET A – Biology: Paper 2 Higher Tier

Author: Mike Smith

Materials

Time allowed: 1 hour 15 minutes

For this paper you must have:
- a ruler
- a calculator

Instructions

- Answer **all** questions in the spaces provided.
- Do all rough work in this book. Cross through any work you do not want to be marked.

Information

- There are 70 marks available on this paper.
- The marks for questions are shown in brackets.
- You are expected to use a calculator where appropriate.
- You are reminded of the need for good English and clear presentation in your answers.
- When answering questions 01.4, 02.1 and 07.1 you need to make sure that your answer:
  – is clear, logical, sensibly structured
  – fully meets the requirements of the question
  – shows that each separate point or step supports the overall answer.

Advice

- In all calculations, show clearly how you work out your answer.

Name:

01 In a park, some grassland is left to grow wild except for a path, which is mown regularly.

Students used a transect line to investigate how the path affected the distribution of four different plant species.

Figure 1.1 shows the line of the transect.

The students placed quadrats every metre along the transect.

Figure 1.1

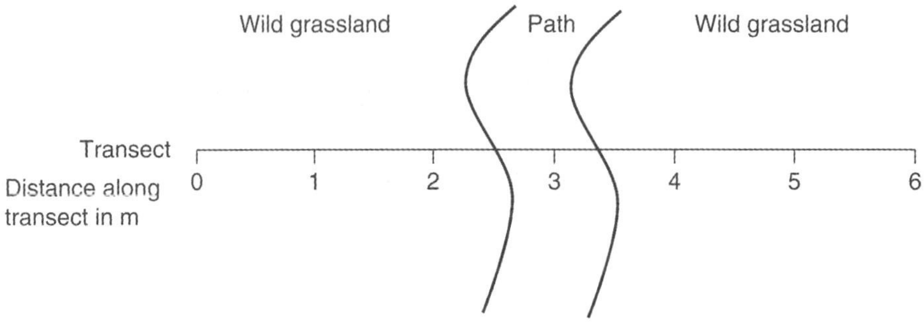

Table 1.1 shows their results.

Table 1.1

Distance along transect in m		0	1	2	3	4	5	6
Number of individual plants of each species per quadrat	**Species A**	5	4	3	0	4	6	5
	Species B	0	0	1	8	2	0	0
	Species C	4	3	2	0	3	4	4
	Species D	0	0	2	3	1	0	0

01.1 Look at **Table 1.1**.

What is the mode number per quadrat for species D?

Answer: _____

[1 mark]

01.2 Look at **Table 1.1**.

What is the median number per quadrat for species A?

Answer: _____

[1 mark]

Question 1 continues on the next page

01.3 **Figure 1.2** shows kite diagrams of the results.

Use the data for **species A** from **Table 1.1** to complete **Figure 1.2**.

Figure 1.2

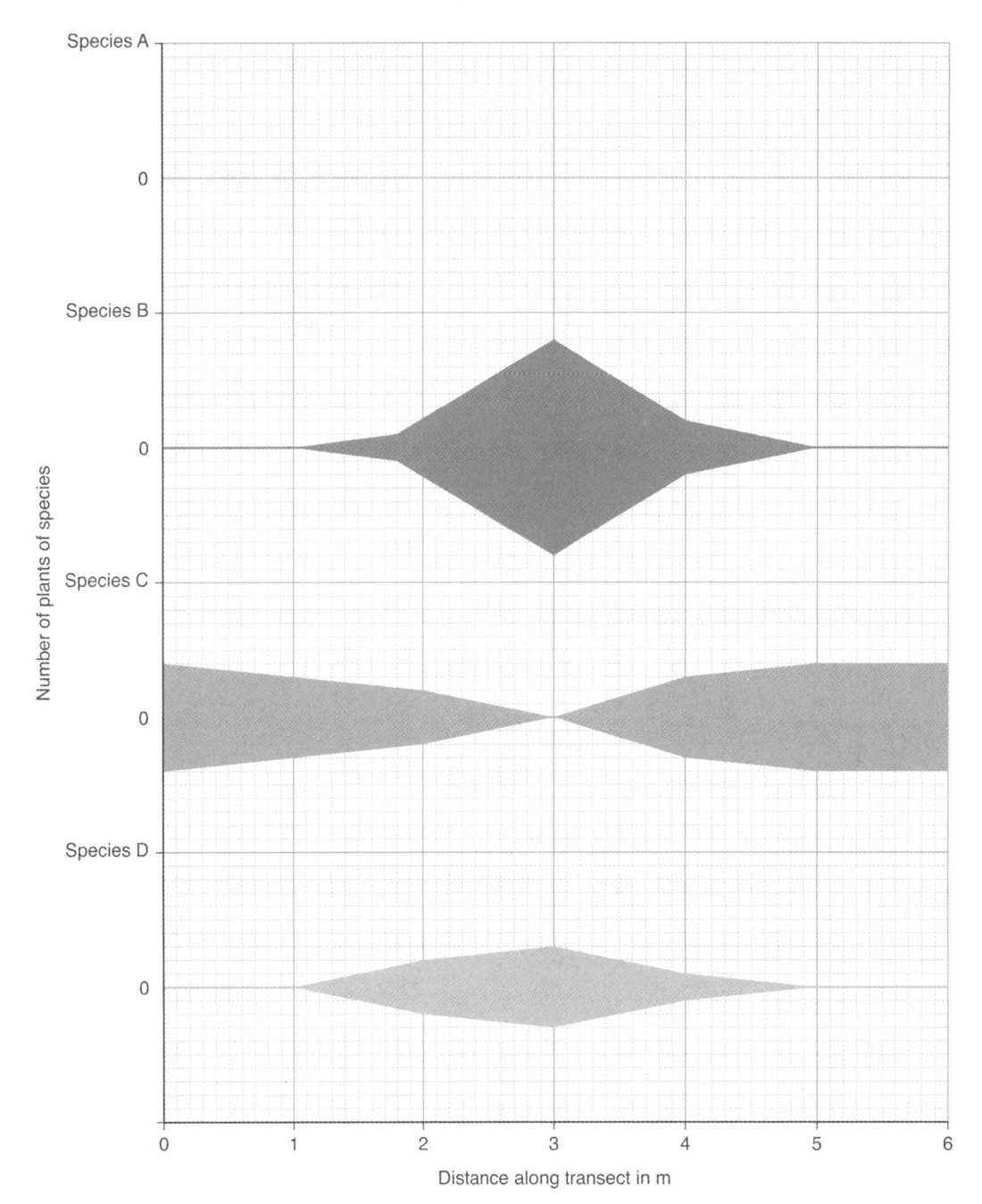

[4 marks]

01.4 **Figure 1.3** shows pictures of each plant species.

Figure 1.3

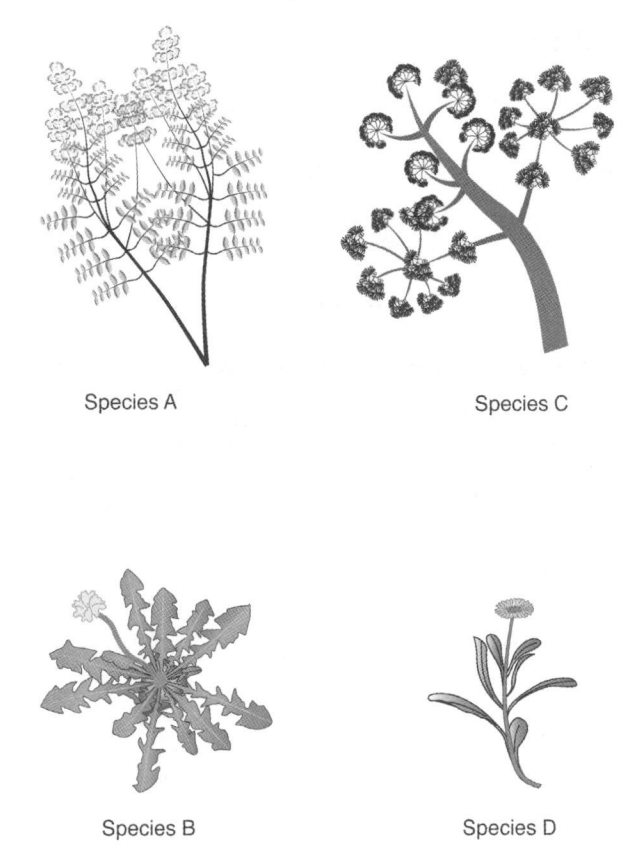

Species A

Species C

Species B

Species D

Suggest reasons for the distributions of the four species along the transect.

Use information from **Table 1.1** and **Figures 1.1, 1.2** and **1.3** to help you answer.

[4 marks]

Turn over >

02 A group of students investigated their reaction times.

They each took it in turn to press a timer button as soon as they heard a buzzer.

Each student used their right hand.

Each student took the test three times and recorded their shortest reaction time.

There were eight girls and six boys in the group.

Table 2.1 shows their results.

Table 2.1

	Shortest reaction times in s								Mean reaction time in s
Girls	0.21	0.16	0.18	0.19	0.18	0.16	0.20	0.19	0.18
Boys	0.19	0.15	0.32	0.16	0.17	0.20			0.20

02.1 One of the students made this conclusion:

Girls have shorter reaction times than boys.

Evaluate the method used and the student's conclusion.

[6 marks]

02.2 **Figure 2.1** shows the nerve pathway involved in the investigation.

Figure 2.1

Sound of buzzer ⟶ Ear ⟶ Brain ⟶ Hand muscles ⟶ Press button

In **Figure 2.1**, which is the receptor and which is the effector?

Receptor: _____

Effector: _____

[2 marks]

02.3 How does information pass along a nerve pathway?

_____

_____

[2 marks]

Question 2 continues on the next page

02.4 One of the students says:

Pressing the button quickly is an example of a reflex action.

Is the student correct?

Give a reason for your answer.

Is the student correct? _____

Reason: _____

[1 mark]

03 Cells can divide by mitosis or meiosis.

03.1 **Table 3.1** shows some features of mitosis or meiosis.

Complete **Table 3.1** by putting a tick (✓) or cross (✗) in each of the empty boxes.

Table 3.1

	Mitosis	**Meiosis**
Involved in body growth		
New cells produced have two copies of each chromosome		
Produces gametes		
Produces genetically identical cells		

[2 marks]

03.2 **Figure 3.1** shows some of the stages of a cell dividing by meiosis.

Write numbers **1**, **2**, **3** and **4** in the boxes to show the correct sequence.

Figure 3.1

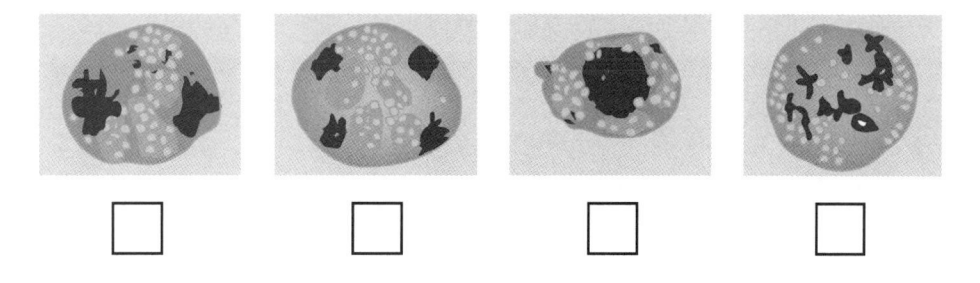

[2 marks]

Question 3 continues on the next page

03.3 Human males have the genotype **XY** and human females have the genotype **XX**.

What is the probability of a couple having a baby girl?

Draw a genetic diagram to explain your answer.

Probability = _____

[4 marks]

03.4 A couple already have one baby girl.

What is the probability that their next baby will also be a girl?

_____

[1 mark]

04 In vitro fertilisation (IVF) includes the following steps.

1. The hormones FSH (follicle stimulating hormone) and LH (luteinising hormone) are given to the mother.
2. Eggs are collected from the mother and sperm from the father.
3. The fertilised eggs develop into embryos.
4. One or more embryos are inserted into the mother's uterus.

04.1 Explain why the mother is given FSH and LH.

[1 mark]

04.2 Where does fertilisation take place?

[1 mark]

04.3 Often, several embryos are inserted into the mother.

Explain why more than one embryo might be inserted.

Describe the possible disadvantage of inserting several embryos.

Reason:

Disadvantage:

[4 marks]

Question 4 continues on the next page

04.4 **Figure 4.1** shows how the levels of four hormones vary during the menstrual cycle.

Figure 4.1

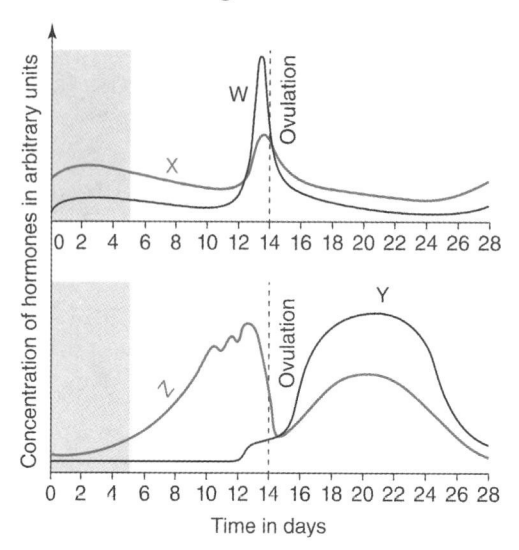

Write down the correct letter for each hormone.

FSH = ...

LH = ...

Oestrogen = ...

Progesterone = ...

[3 marks]

05 Polydactyly is a condition in which a person has extra fingers or toes.

It is an inherited disorder caused by a dominant allele **D**.

The recessive allele is **d**.

05.1 What is the genotype of someone who is heterozygous for polydactyly?

[1 mark]

05.2 What is the genotype of someone who is homozygous dominant for polydactyly?

[1 mark]

05.3 What is the phenotype of someone with the genotype **dd**?

[1 mark]

05.4 Is it possible for two parents who do **not** have the polydactyly condition to have a child with the condition?

Explain your answer.

Answer:

Explanation:

[2 marks]

Question 5 continues on the next page

05.5 **Figure 5.1** shows a family tree in which polydactyly occurs.

Figure 5.1

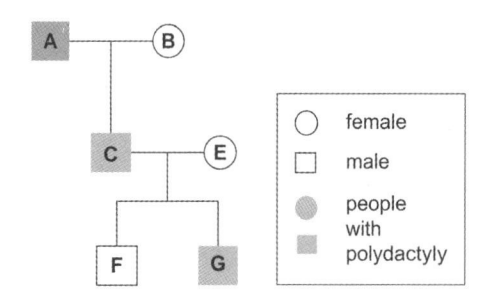

What are the possible genotypes of **A** and **C**?

Give reasons for your answers.

A = ...

C = ...

Reasons: ....

....

....

....

....

[4 marks]

05.6 Unlike polydactyly, most other inherited disorders are caused by **recessive** alleles.

Suggest why inherited disorders are more commonly caused by recessive alleles.

....

....

....

[1 mark]

06 **Figure 6.1** shows a type of bird called a St Kilda wren.

Figure 6.1

St Kilda wrens live on the island of St Kilda off the north coast of Scotland.

They are similar to wrens that live on the mainland, but St Kilda wrens are larger.

Wrens are too small to normally fly to or from the island.

Scientists think that:

- St Kilda wrens are descended from mainland wrens that were blown over to the island by strong winds
- their larger size is an adaptation to help keep warm.

Question 6 continues on the next page

06.1 Scientists think that the St Kilda wrens have evolved from mainland wrens by natural selection.

Describe how this may have happened.

[4 marks]

06.2 Although St Kilda wrens are different from the mainland wrens they are classified as the **same** species.

Describe how you could show that the mainland wrens and the St Kilda wrens are the same species.

[2 marks]

06.3 The mainland wrens have the scientific name *Troglodytes troglodytes*.

The St Kilda wrens have the scientific name *Troglodytes troglodytes hirtensis*.

Suggest why the St Kilda wrens have this different scientific name.

....

....

....

[2 marks]

06.4 In the future, the St Kilda wrens may evolve to become so different from the mainland wrens that they could be classified as a different species.

Which of the following could be an appropriate name for the new species?

Tick **one** box and give a reason for your answer.

Hirtensis hirtensis ☐

Hirtensis troglodytes ☐

Troglodytes hirtensis ☐

Troglodytes troglodytes ☐

Reason: ....

....

....

[2 marks]

Turn over >

07 Bylot Island is in the Arctic.

Not many animal species live on Bylot Island.

Two species that do live there are snowy owls and lemmings, as shown in **Figure 7.1**.

Snowy owls catch and eat lemmings.

Figure 7.1

Snowy owl   Lemming

07.1 **Figure 7.2** shows data from Bylot Island.

Figure 7.2

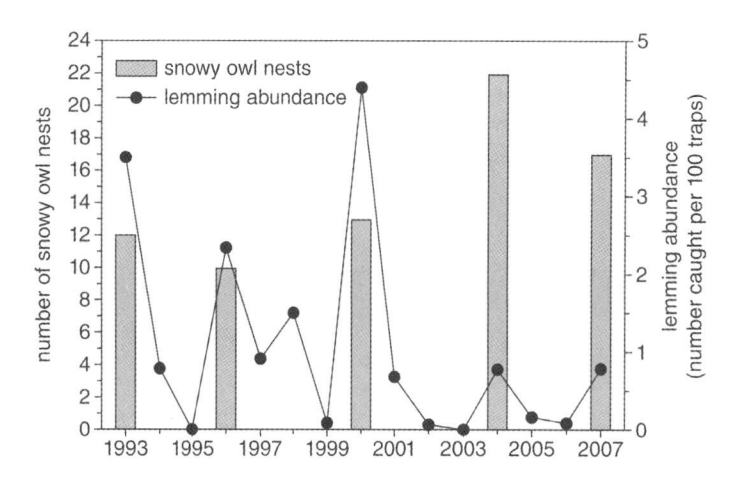

Describe any patterns in the data shown in **Figure 7.2**.

Suggest explanations for these patterns.

[6 marks]

07.2 Snowy owl bodies contain carbon.

Eventually this carbon is recycled back into the atmosphere as carbon dioxide.

Describe how this happens.

[3 marks]

Question 7 continues on the next page

07.3 Bylot Island has a low biodiversity.

Student A says:

It is important to protect places like Bylot Island.

Student **B** says:

It is more important to protect places with a higher biodiversity like tropical rainforest.

Evaluate these two statements.

[2 marks]

END OF QUESTIONS

AQA
GCSE
Combined Science: Trilogy H
SET A – Chemistry: Paper 3 Higher Tier

Author: Sunetra Berry

Materials

Time allowed: 1 hour 15 minutes

For this paper you must have:
- a ruler
- a calculator
- the Periodic Table (found at the end of the paper).

Instructions

- Answer **all** questions in the spaces provided.
- Do all rough work in this book. Cross through any work you do not want to be marked.

Information

- There are 70 marks available on this paper.
- The marks for questions are shown in brackets.
- You are expected to use a calculator where appropriate.
- You are reminded of the need for good English and clear presentation in your answers.
- When answering question 05.1 you need to make sure that your answer:
  – is clear, logical, sensibly structured
  – fully meets the requirements of the question
  – shows that each separate point or step supports the overall answer.

Advice

- In all calculations, show clearly how you work out your answer.

Name: _____

01 This question is about bonding and structure.

    01.1 Which of the following structures does **not** have weak intermolecular forces?

    Tick **one** box.

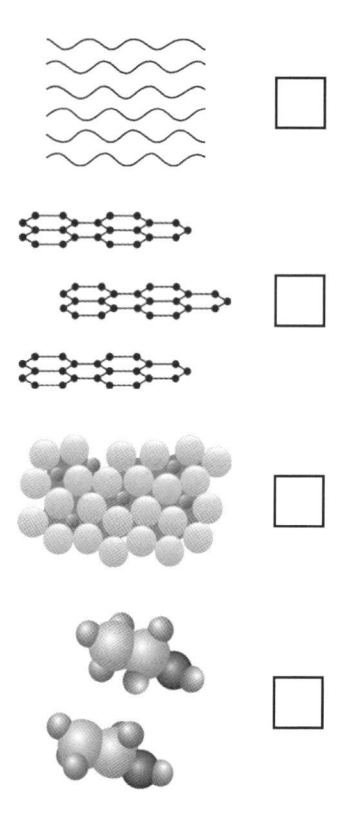

☐

☐

☐

☐

[1 mark]

01.2 The formation of sodium chloride, NaCl, can be represented by this equation.

$$Na\bullet + {}^{\times}_{\times}\overset{\times\times}{Cl}{}^{\times}_{\times} \longrightarrow [Na]^{+} \; [{}^{\times}_{\times}\overset{\times\times}{Cl}{}^{\times}_{\times}]^{-}$$

(2, 8, 1) (2, 8, 7)   (2, 8) (2, 8, 8)

Write a similar equation to show the formation of magnesium fluoride, MgF$_2$.

[2 marks]

01.3 Which one of the following structures of carbon does **not** conduct electricity?

Tick **one** box.

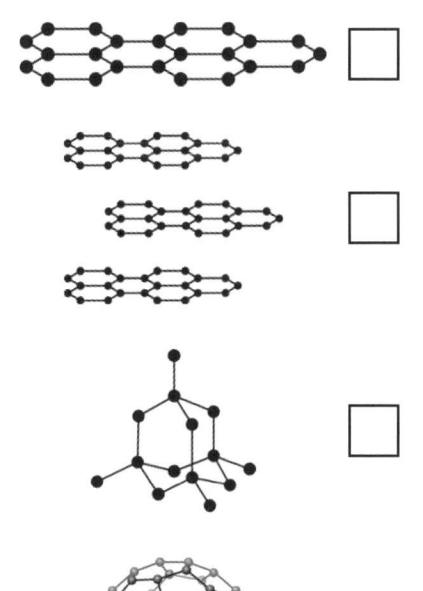

[1 mark]

Question 1 continues on the next page

01.4 List the main differences between **small** covalent compounds and **giant** covalent compounds, and their properties.

[3 marks]

02

02.1 Which of the elements in Group 7 has the highest boiling **and** melting points?

Tick **one** box.

Bromine ☐

Chlorine ☐

Fluorine ☐

Iodine ☐

[1 mark]

02.2 Draw a dot and cross diagram to show the bonding in a fluorine molecule.

The atomic number for fluorine is 9

Show the outermost electrons only.

[2 marks]

Question 2 continues on the next page

02.3 Group 1 elements react with water.

Describe:

- observations you would see in reactions between Group 1 elements and water
- the way the reactivity changes down Group 1.

Group 7 elements react with dissolved salts of less reactive halogens.

Describe:

- observations you would see in reactions between Group 7 elements and dissolved salts of less reactive halogens
- the way the reactivity changes down Group 7.

[4 marks]

03

03.1 The electrolysis cell shown in **Figure 3.1** contains an electrolyte of molten sodium iodide.

The two electrodes are unreactive.

Figure 3.1

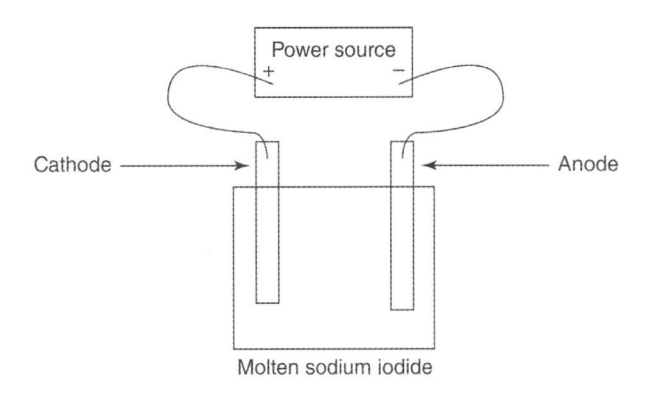

Which statement is **true**?

Tick **one** box.

Sodium is formed at the anode. ☐

Iodine is formed at the anode. ☐

Hydrogen is formed at the cathode. ☐

Hydrogen is formed at the anode. ☐

[1 mark]

Question 3 continues on the next page

03.2 Explain why magnesium, sodium and potassium must be extracted from their ores using electrolysis, and not using other methods.

[2 marks]

03.3 Describe what you would observe in the electrolysis of **aqueous copper chloride**.

Explain the observations.

Describe which products would be formed at each electrode.

[4 marks]

04

04.1 The thermal decomposition of calcium carbonate occurs at about 700 °C

Draw a reaction profile for this reaction using the axes below.

On your profile, label:

- the relative energies of the reactants and products
- the activation energy
- the overall energy change.

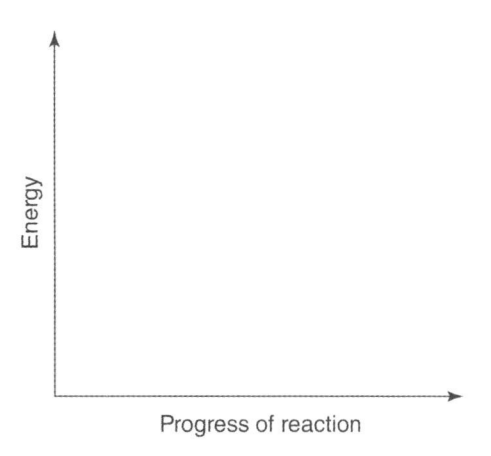

[Axes: Energy (y-axis) vs Progress of reaction (x-axis)]

[2 marks]

04.2 A different reaction has the same type of energy change as the reaction above, but it takes place at room temperature.

In what way will the reaction profile be different to the profile above?

....

....

[1 mark]

Question 4 continues on the next page

04.3 **Table 4.1** shows some bond energy data.

Table 4.1

Bond	kJ/mol
C–H	414
C–C	347
C–Br	276
C=C	615
H–Br	368

Calculate the overall energy change for this complete reaction:

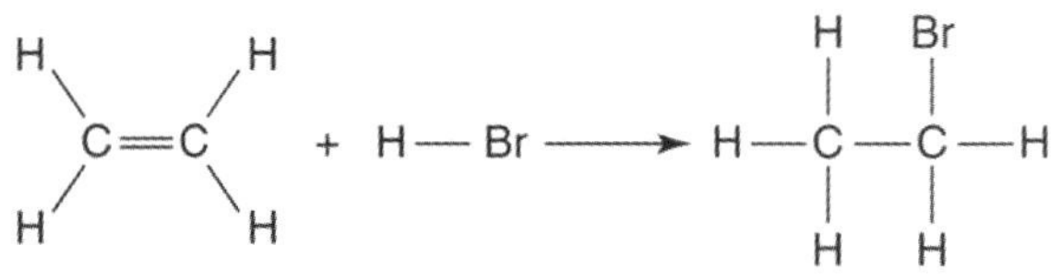

$$CH_2=CH_2 + H-Br \rightarrow CH_3-CH_2Br$$

Use the data given in **Table 4.1**.

Overall energy change = ... kJ/mol

[4 marks]

05

05.1 A student wanted to determine the reactivity of two unknown metals, X and Y.

She reacted each metal with salt solutions of known metals and recorded her observations and temperature changes.

Table 5.1 shows her results.

Table 5.1

Salt solution	Observations	Initial temp (°C)	Final temp (°C)
Reactions with metal X:			
copper sulfate (blue solution)	brown layer formed on metal	20	32
magnesium sulfate	no reaction observed	20	20
iron(II) sulfate (green solution)	no reaction observed	20	25
Reactions with metal Y:			
copper sulfate (blue solution)	brown layer formed on metal and blue solution becomes colourless	20	28
magnesium sulfate	no reaction observed	20	20
iron(II) sulfate (green solution)	dark crystals observed and green solution becomes colourless	20	23

Question 5 continues on the next page

Describe how the experiment might be carried out to obtain the data in **Table 5.1**.

Describe:

- which measurements should be made, and how they should be made
- how to make the experiment fair
- how to obtain reliable, precise data.

[6 marks]

05.2 Draw **three** conclusions about the reactivity of the metals.

Use the data in **Table 5.1**.

[3 marks]

05.3 Does the data in **Table 5.1** provide enough information to determine which metal, X or Y, is most reactive?

Explain your answer.

[2 marks]

05.4 Suggest one further test which could be performed to determine which metal, X or Y, is most reactive.

[2 marks]

Turn over >

06 A student wants to make a pure solution of iron(II) sulfate from iron(III) oxide.

06.1 At first, he roasts the iron(III) oxide with carbon to produce iron and carbon dioxide.

Write a **balanced** symbol equation for the reaction.

Use your Periodic Table to help you.

[2 marks]

06.2 What is the name for this type of reaction?

[1 mark]

06.3 The student reacts the iron he has made with sulfuric acid in the following reaction:

$$Fe(s) + H_2SO_4(aq) \longrightarrow FeSO_4(aq) + H_2(g)$$

Calculate the mass of iron needed to react with 50 cm³ of 2 mol/dm³ of sulfuric acid.

Show your working.

(M_r of Fe = 56)

Mass of iron = _____ g

[3 marks]

06.4 The student has only 4 g of iron.

What is the maximum mass of iron(II) sulfate he can make?

Show your working.

(M_r of Fe = 56; M_r of FeSO$_4$ = 150)

Mass of iron(II) sulfate = ... g

[3 marks]

06.5 Explain the term **limiting reactant** using the reaction between iron and sulfuric acid.

[2 marks]

06.6 The student dissolves 5 g of the iron(II) sulfate to make a solution with a concentration of 0.33 mol/dm³.

Calculate the volume of liquid required to make up this concentration.

Show your working.

[3 marks]

06.7 Sometimes iron forms an alloy with carbon.

Describe the differences in structure between a **pure metal** and an **alloy**.

Explain why this affects their properties.

[3 marks]

07 In an experiment:

- 60 g of ethanoic acid (CH$_3$COOH) is dissolved in water to make 1 dm^3 of solution
- 12.16 g of hydrochloric acid (HCl) is also dissolved in water to make 1 dm^3 of solution.

07.1 Calculate the concentration of each acid in mol/dm^3

Give your answers to 2 significant figures.

(M_r of CH$_3$COOH = 46; M_r of HCl = 36.5)

[2 marks]

07.2 Explain why the pH of the HCl solution is lower than the pH of the ethanoic acid solution.

[2 marks]

Question 7 continues on the next page

07.3 Sulfuric acid reacts with magnesium to make a solution of magnesium sulfate.

Which species is oxidised and which is reduced in terms of electron transfer?

Oxidised: ....

Reduced: ....

[2 marks]

07.4 Write a full ionic equation for the reaction.

....

....

[2 marks]

07.5 Draw and label an electrolytic cell to electrolyse magnesium sulfate solution.

For each electrode:

- describe the reaction that takes place, and what would be observed
- explain why this reaction takes place, using half equations.

[4 marks]

END OF QUESTIONS

The Periodic Table

Key
- Metals
- Non-metals

Key entry format:
- Relative atomic mass → 1
- Atomic symbol → H
- Name → hydrogen
- Atomic/proton number → 1

1	2											3	4	5	6	7	0 or 8
																	4 **He** helium 2
7 **Li** lithium 3	9 **Be** beryllium 4											11 **B** boron 5	12 **C** carbon 6	14 **N** nitrogen 7	16 **O** oxygen 8	19 **F** fluorine 9	20 **Ne** neon 10
23 **Na** sodium 11	24 **Mg** magnesium 12											27 **Al** aluminium 13	28 **Si** silicon 14	31 **P** phosphorus 15	32 **S** sulfur 16	35.5 **Cl** chlorine 17	40 **Ar** argon 18
39 **K** potassium 19	40 **Ca** calcium 20	45 **Sc** scandium 21	48 **Ti** titanium 22	51 **V** vanadium 23	52 **Cr** chromium 24	55 **Mn** manganese 25	56 **Fe** iron 26	59 **Co** cobalt 27	59 **Ni** nickel 28	63.5 **Cu** copper 29	65 **Zn** zinc 30	70 **Ga** gallium 31	73 **Ge** germanium 32	75 **As** arsenic 33	79 **Se** selenium 34	80 **Br** bromine 35	84 **Kr** krypton 36
85 **Rb** rubidium 37	88 **Sr** strontium 38	89 **Y** yttrium 39	91 **Zr** zirconium 40	93 **Nb** niobium 41	96 **Mo** molybdenum 42	[98] **Tc** technetium 43	101 **Ru** ruthenium 44	103 **Rh** rhodium 45	106 **Pd** palladium 46	108 **Ag** silver 47	112 **Cd** cadmium 48	115 **In** indium 49	119 **Sn** tin 50	122 **Sb** antimony 51	128 **Te** tellurium 52	127 **I** iodine 53	131 **Xe** xenon 54
133 **Cs** caesium 55	137 **Ba** barium 56	139 **La*** lanthanum 57	178 **Hf** hafnium 72	181 **Ta** tantalum 73	184 **W** tungsten 74	186 **Re** rhenium 75	190 **Os** osmium 76	192 **Ir** iridium 77	195 **Pt** platinum 78	197 **Au** gold 79	201 **Hg** mercury 80	204 **Tl** thallium 81	207 **Pb** lead 82	209 **Bi** bismuth 83	[209] **Po** polonium 84	[210] **At** astatine 85	[222] **Rn** radon 86
[223] **Fr** francium 87	[226] **Ra** radium 88	[227] **Ac*** actinium 89	[261] **Rf** rutherfordium 104	[262] **Db** dubnium 105	[266] **Sg** seaborgium 106	[264] **Bh** bohrium 107	[277] **Hs** hassium 108	[268] **Mt** meitnerium 109	[271] **Ds** darmstadtium 110	[272] **Rg** roentgenium 111	[285] **Cn** copernicium 112	[286] **Uut** ununtrium 113	[289] **Fl** flerovium 114	[289] **Uup** ununpentium 115	[293] **Lv** livermorium 116	[294] **Uus** ununseptium 117	[294] **Uuo** ununoctium 118

*The lanthanides (atomic numbers 58–71) and the actinides (atomic numbers 90–103) have been omitted.
The relative atomic masses of copper and chlorine have not been rounded to the nearest whole number.

AQA
GCSE
Combined Science: Trilogy H
SET A – Chemistry: Paper 4 Higher Tier

Author: Sunetra Berry

Materials

Time allowed: 1 hour 15 minutes

For this paper you must have:
- a ruler
- a calculator
- the Periodic Table (found at the end of the paper).

Instructions

- Answer **all** questions in the spaces provided.
- Do all rough work in this book. Cross through any work you do not want to be marked.

Information

- There are 70 marks available on this paper.
- The marks for questions are shown in brackets.
- You are expected to use a calculator where appropriate.
- You are reminded of the need for good English and clear presentation in your answers.
- When answering questions 05.1 and 07.1 you need to make sure that your answer:
  – is clear, logical, sensibly structured
  – fully meets the requirements of the question
  – shows that each separate point or step supports the overall answer.

Advice

- In all calculations, show clearly how you work out your answer.

Name: _____

01

01.1 Describe the differences between **potable** water and **pure** water.

[2 marks]

01.2 Sterilisation is used to try to remove all pathogens from water.

Figure 1.1 shows the results of a study which compares three main sterilising agents: chlorine, ozone and ultraviolet.

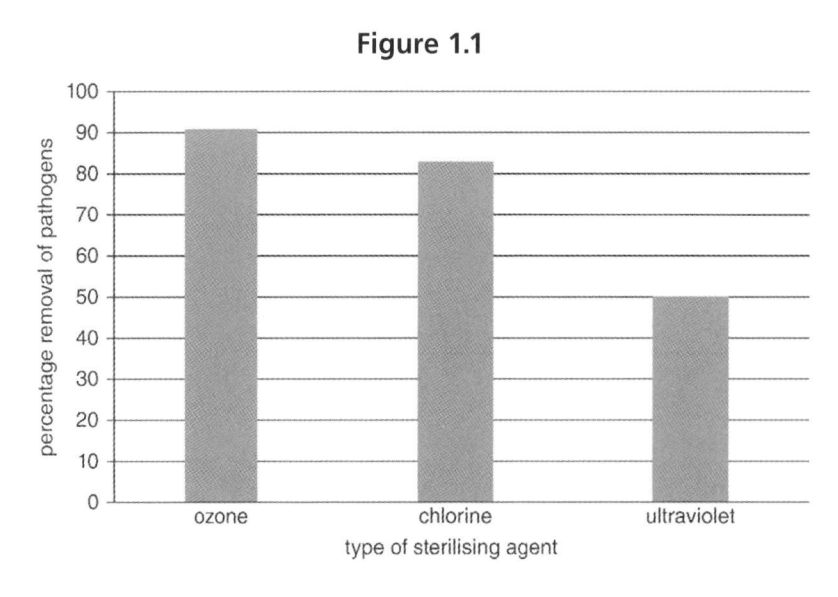

Figure 1.1

What **two** conclusions can you draw from these results?

Use data from the graph to support your answers.

1.

2.

[2 marks]

01.3 State **two** ways that the investigators could make the study a fair test.

1. _____

2. _____

[2 marks]

01.4 What is different about the **treatment of sewage,** when compared with the **treatment of ground water?**

_____

_____

_____

_____

_____

_____

_____

[4 marks]

02

02.1 A student investigated the rate of reaction between magnesium and an acid.

The student used magnesium powder and magnesium ribbon.

The results are presented in **Figure 2.1**, but the student forgot to label the data.

Draw a line of best fit for each set of data.

Label each line of best fit as either **powder** or **ribbon**.

Figure 2.1

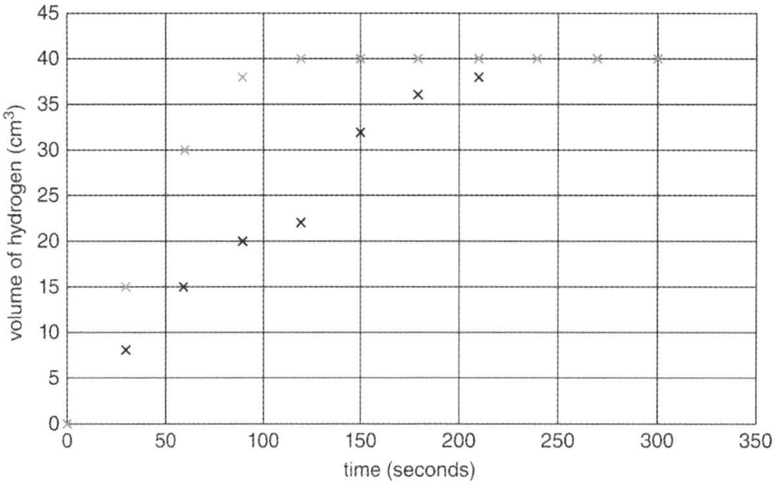

[3 marks]

02.2 Calculate the mean rate of reaction for the **powder**.

Use data from the graph.

Give your answer to 3 significant figures.

Rate = ... cm^3/s

[3 marks]

02.3 Explain why there is a difference in the rate of reaction with powder and ribbon.

Use the idea of particles in your explanation.

[2 marks]

02.4 The temperature of the acid is now increased by 10 °C

Explain why this will affect the rate of **both** reactions.

Use the idea of particles in your explanation.

[3 marks]

03

03.1 Which statement about alkanes is **true**?

Tick **one** box.

The general formula for an alkane is C_nH_{2n+2}. ☐

Alkanes only exist as mixtures of hydrocarbons. ☐

Alkanes and alkenes are chemically identical. ☐

Alkanes are very reactive. ☐

[1 mark]

03.2 Crude oil is a mixture which is separated using fractional distillation.

Table 3.1 gives some information about the fractions produced using fractional distillation.

The fractions are not in the correct order.

Table 3.1

Fraction	Number of carbon atoms	Boiling point (°C)
kerosene	10–16	
fuel oil	20–79	
petrol		40–80
diesel oil	16–20	250–350
liquefied petroleum gas		20–40

Some of the data is missing.

Choose from the following values and add them to **Table 3.1**.

| above 350 | 150–250 | 5–10 | 80–100 | 1–4 |

[4 marks]

03.3 Write a balanced symbol equation for the complete combustion of butane.

[2 marks]

03.4 Describe **three** differences between the properties of methane and decane ($C_{10}H_{22}$).

[3 marks]

03.5 Hydrocarbons can be made into new products by the processes of **cracking** and **polymerisation**.

Describe the difference between cracking and polymerisation.

[4 marks]

Turn over >

04

04.1 Ibuprofen is a common painkiller. It is usually bought at a pharmacy in tablet form. Each tablet is a formulation.

Which of the following statements about one tablet of ibuprofen is **true**?

Tick **one** box.

It consists only of ibuprofen. ☐

It has a fixed melting point. ☐

It is a carefully controlled mixture. ☐

It is a single compound. ☐

[1 mark]

04.2 The structure of ibuprofen is shown in **Figure 4.1**.

Figure 4.1

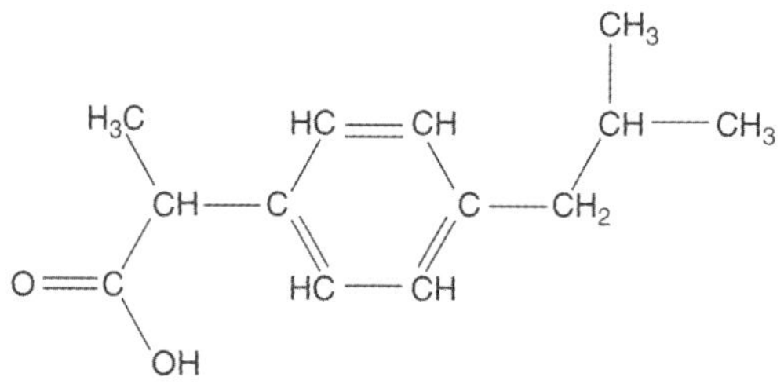

What is the molecular formula of ibuprofen?

Calculate the relative formula mass. (A_r values: C = 12; H = 1; O = 16)

[2 marks]

04.3 Chromatography can be used to detect the purity of different medicines, such as antibiotics.

Describe how chromatography works.

[3 marks]

04.4 **Figure 4.2** shows the results from testing from different antibiotic tablets. Each tablet contains either one or two active ingredients (ingredients that have antibiotic properties).

Which tablet consists of the most ingredients?

Explain your answer.

Figure 4.2

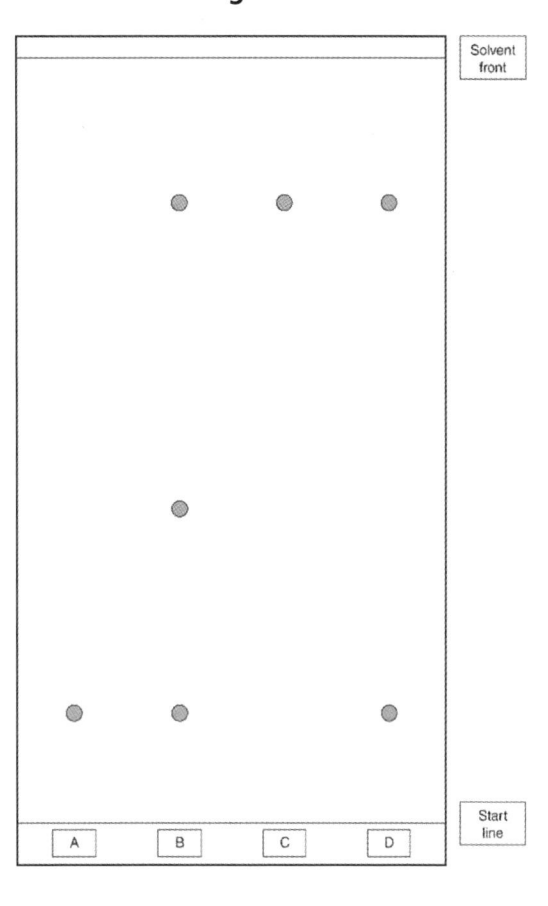

[A chromatogram showing solvent front at top and start line at bottom. Four lanes labelled A, B, C, D. Lane A has one spot near the top. Lane B has two spots: one near the top and one in the middle. Lane C has one spot near the top. Lane D has one spot near the bottom.]

....

....

[1 mark]

Question 4 continues on the next page

04.5 Draw **three** conclusions about the tablets from the chromatogram.

[3 marks]

04.6 Calculate the R_f value of antibiotic A.

Use **Figure 4.2**.

Show your working.

Give your answer to 2 significant figures.

[3 marks]

05

05.1 Copper is usually obtained from ores with a high copper content.

It can also be obtained from low grade ores by other methods.

Compare **two** different methods of obtaining copper from low grade ores.

Outline the advantages and disadvantages of **each** method.

[6 marks]

06 Carbon dioxide and methane are two greenhouse gases.

06.1 State **two** human activities that can cause methane to be present in the atmosphere.

[2 marks]

06.2 Explain how these gases cause the greenhouse effect.

[2 marks]

06.3 Look at Graph A and Graph B in **Figure 6.1**.

Figure 6.1

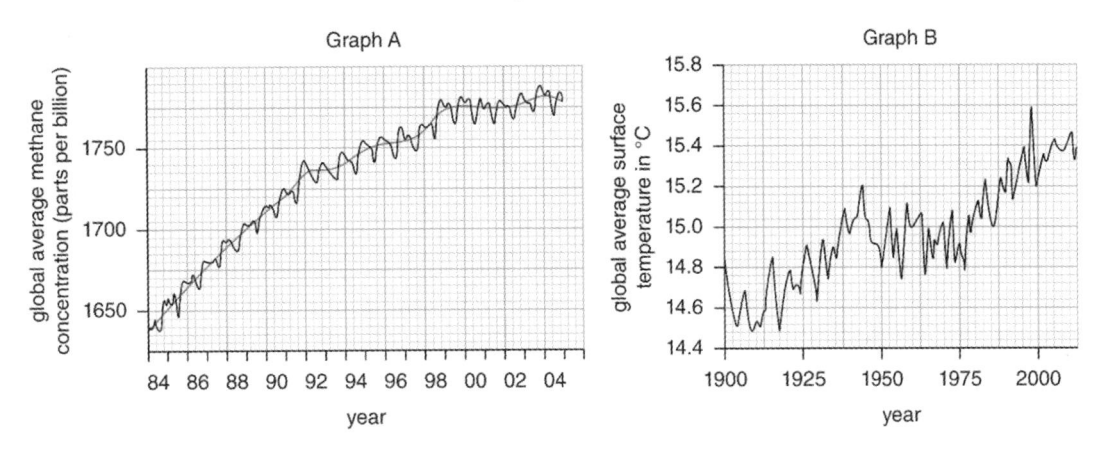

Graph A / Graph B (global average methane concentration in parts per billion, 1984–2004; global average surface temperature in °C, 1900–2000+)

What evidence does this data provide that methane contributes to global warming?

Select values from the graphs to support your ideas.

[2 marks]

Question 6 continues on the next page

06.4 Why is it difficult to draw **definite** conclusions using this data alone?

Give **three** reasons.

[3 marks]

06.5 Describe the chemical test for carbon dioxide and the expected result.

[1 mark]

07

07.1 Methane is used to produce hydrogen gas.

The first step of the process is given here.

$$CH_4(g) + H_2O(g) \rightleftharpoons CO(g) + 3H_2(g)$$

Energy change = +206 kJ/mol

The reaction takes place at 700 °C

It is an equilibrium reaction and, in the forward reaction, energy is taken in.

Describe what would happen to the equilibrium **and** the yield of hydrogen if:

- the temperature was decreased from 700 °C to 500 °C
- the pressure was increased.

[6 marks]

END OF QUESTIONS

The Periodic Table

1	2											3	4	5	6	7	0 or 8
																	4 **He** helium 2
7 **Li** lithium 3	9 **Be** beryllium 4											11 **B** boron 5	12 **C** carbon 6	14 **N** nitrogen 7	16 **O** oxygen 8	19 **F** fluorine 9	20 **Ne** neon 10
23 **Na** sodium 11	24 **Mg** magnesium 12											27 **Al** aluminium 13	28 **Si** silicon 14	31 **P** phosphorus 15	32 **S** sulfur 16	35.5 **Cl** chlorine 17	40 **Ar** argon 18
39 **K** potassium 19	40 **Ca** calcium 20	45 **Sc** scandium 21	48 **Ti** titanium 22	51 **V** vanadium 23	52 **Cr** chromium 24	55 **Mn** manganese 25	56 **Fe** iron 26	59 **Co** cobalt 27	59 **Ni** nickel 28	63.5 **Cu** copper 29	65 **Zn** zinc 30	70 **Ga** gallium 31	73 **Ge** germanium 32	75 **As** arsenic 33	79 **Se** selenium 34	80 **Br** bromine 35	84 **Kr** krypton 36
85 **Rb** rubidium 37	88 **Sr** strontium 38	89 **Y** yttrium 39	91 **Zr** zirconium 40	93 **Nb** niobium 41	96 **Mo** molybdenum 42	[98] **Tc** technetium 43	101 **Ru** ruthenium 44	103 **Rh** rhodium 45	106 **Pd** palladium 46	108 **Ag** silver 47	112 **Cd** cadmium 48	115 **In** indium 49	119 **Sn** tin 50	122 **Sb** antimony 51	128 **Te** tellurium 52	127 **I** iodine 53	131 **Xe** xenon 54
133 **Cs** caesium 55	137 **Ba** barium 56	139 **La*** lanthanum 57	178 **Hf** hafnium 72	181 **Ta** tantalum 73	184 **W** tungsten 74	186 **Re** rhenium 75	190 **Os** osmium 76	192 **Ir** iridium 77	195 **Pt** platinum 78	197 **Au** gold 79	201 **Hg** mercury 80	204 **Tl** thallium 81	207 **Pb** lead 82	209 **Bi** bismuth 83	[209] **Po** polonium 84	[210] **At** astatine 85	[222] **Rn** radon 86
[223] **Fr** francium 87	[226] **Ra** radium 88	[227] **Ac*** actinium 89	[261] **Rf** rutherfordium 104	[262] **Db** dubnium 105	[266] **Sg** seaborgium 106	[264] **Bh** bohrium 107	[277] **Hs** hassium 108	[268] **Mt** meitnerium 109	[271] **Ds** darmstadtium 110	[272] **Rg** roentgenium 111	[285] **Cn** copernicium 112	[286] **Uut** ununtrium 113	[289] **Fl** flerovium 114	[289] **Uup** ununpentium 115	[293] **Lv** livermorium 116	[294] **Uus** ununseptium 117	[294] **Uuo** ununoctium 118

Key:
- Metals
- Non-metals

Relative atomic mass → 1
Atomic symbol → **H**
Name → hydrogen
Atomic/proton number → 1

*The lanthanides (atomic numbers 58–71) and the actinides (atomic numbers 90–103) have been omitted.
The relative atomic masses of copper and chlorine have not been rounded to the nearest whole number.

Collins

AQA
GCSE
Combined Science: Trilogy H
SET A – Physics: Paper 5 Higher Tier
Author: Lynn Pharoah

Time allowed: 1 hour 15 minutes

Materials

For this paper you must have:
- a ruler
- a calculator
- the Physics Equation Sheet (found at the end of the paper).

Instructions

- Answer **all** questions in the spaces provided.
- Do all rough work in this book. Cross through any work you do not want to be marked.

Information

- There are 70 marks available on this paper.
- The marks for questions are shown in brackets.
- You are expected to use a calculator where appropriate.
- You are reminded of the need for good English and clear presentation in your answers.
- When answering questions 03.3, 05.1 and 07.1 you need to make sure that your answer:
  – is clear, logical, sensibly structured
  – fully meets the requirements of the question
  – shows that each separate point or step supports the overall answer.

Advice

- In all calculations, show clearly how you work out your answer.

Name: ....

01

01.1 Explain what is meant by **radioactive decay**.

[3 marks]

01.2 Which type of radiation **has a range of a few metres in air**?

Tick **one** box.

Alpha ☐

Beta ☐

Gamma ☐

[1 mark]

01.3 Alpha radiation is described as having the highest **ionising power**.

Describe what is meant by **ionising power**.

State how the high ionising power of alpha radiation affects its range through solid material.

[3 marks]

01.4 A carbon-14 nucleus decays by beta emission to form a nucleus of nitrogen.

Complete the decay equation.

$$^{14}_{6}C \rightarrow \, ^{\underline{}}_{\underline{}}N + \, ^{\underline{}}_{\underline{}}e$$

[4 marks]

02 **Figure 2.1** gives information about the conduction of thermal energy through the stone walls of an old cottage.

Figure 2.1

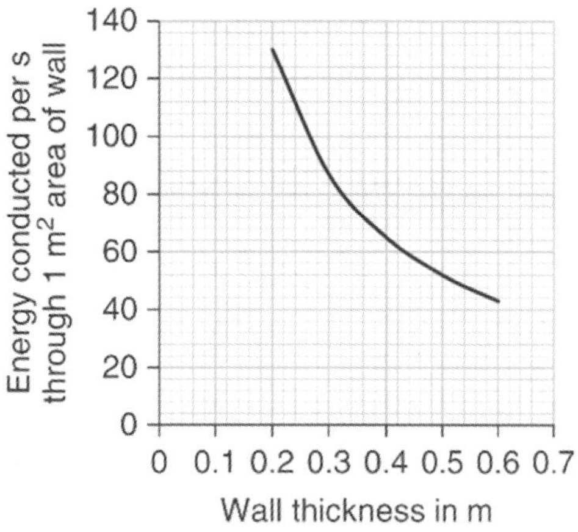

02.1 Use **Figure 2.1** to determine the thermal energy conducted per s through 1 m^2 of stone wall of thickness 0.2 m

Thermal energy conducted = ... W/m^2

[1 mark]

02.2 Use **Figure 2.1** to determine the thermal energy conducted per s through 1 m^2 of stone wall of thickness 0.4 m

Thermal energy conducted = ... W/m^2

[1 mark]

Question 2 continues on the next page

02.3 Write a conclusion about the rate of conduction of thermal energy through a stone wall.

Use the graph in **Figure 2.1** and your answers to **02.1** and **02.2**.

[2 marks]

02.4 The walls of a modern building have been constructed with an air gap between the inner and outer walls as this helps to reduce unwanted energy losses.

State what this indicates about how the thermal conductivity of brick compares with that of air.

Suggest a further change to the walls that could help reduce the rate at which the building cools.

[2 marks]

03.1 The symbol for the stable isotope of gold is $^{197}_{79}\text{Au}$.

Give the number of neutrons and the number of protons in the nucleus of this isotope.

Number of neutrons = _____

Number of protons = _____

[2 marks]

03.2 Describe the number and location of the electrons in a neutral atom of gold, $^{197}_{79}\text{Au}$.

_____

_____

[2 marks]

03.3 In the alpha particle scattering experiment, a beam of alpha particles was fired at gold foil.

Any change in direction of the alpha particles on hitting the gold foil was monitored.

Table 3.1 shows some typical results.

Table 3.1

Path followed by the alpha particles	Percentage of alpha particles
Undeflected and continued in a straight line	>99%
Deflected by angles up to 90°	0.05%
Deflected by angles greater than 90°	0.013%

Question 3 continues on the next page

Describe how the results of the alpha particle scattering experiment led to the nuclear model of the atom replacing the plum pudding model.

Use the data in **Table 3.1**.

[4 marks]

04

04.1 Define **specific heat capacity**.

[1 mark]

04.2 An electric kettle is used to heat 1.0 kg of water.

Calculate the thermal energy that must be transferred to the water to raise its temperature from 20°C to 100°C

Use the correct equation from the Physics Equation Sheet.

The specific heat capacity of water is 4200 J/kg °C

Thermal energy = _____ J

[2 marks]

04.3 Write down the equation that links power, energy transferred and time.

[1 mark]

04.4 The electrical heater inside the kettle has a power of 3000 W

The kettle takes 120 s to bring the water to its boiling point.

Calculate the thermal energy transferred by the heater in the time taken to boil the water.

Energy transferred = _____ J

[3 marks]

Question 4 continues on the next page

04.5 Write down the equation for energy efficiency.

[1 mark]

04.6 Calculate the efficiency of the kettle.

Give your answer as a percentage to 2 significant figures.

Efficiency = ... %

[3 marks]

04.7 The kettle is not 100% efficient.

This is partly because thermal energy is dissipated to the surrounding room.

Suggest **one other** reason that the kettle is not 100% efficient.

[1 mark]

05

05.1 A student is investigating how the resistance of a piece of wire depends on its length.

The student has the apparatus shown in **Figure 5.1** and a metre rule.

Wires can be connected between crocodile clips X and Y.

Write a series of step-by-step instructions that the student could follow to determine the resistance of a range of different lengths of wire.

Figure 5.1

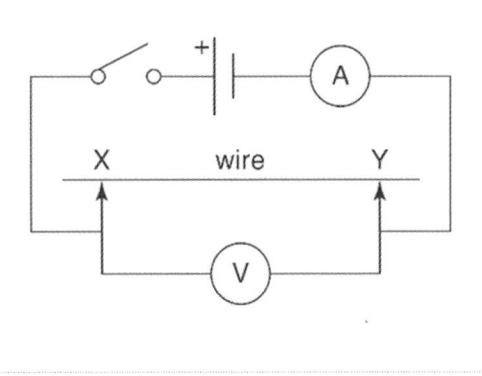

[4 marks]

05.2 The student decides to take repeat measurements for each length of wire.

What type of error does taking repeat measurements help to remove?

[1 mark]

Question 5 continues on the next page

05.3 After the experiment, the student realised that the metre rule had been damaged at one end.

Instead of starting at 0 mm, the end of the rule corresponded to the 1 mm mark.

State what type of error this creates in the measurement of the wire's length.

....

[1 mark]

05.4 Explain how the student should correct the length measurements made in the experiment.

....

....

....

[1 mark]

05.5 The resistance of the student's wire doubles each time the length connected between X and Y is doubled.

Draw a **sketch** graph on the axes in **Figure 5.2** below, to show how the resistance of the wire depends on its length.

Figure 5.2

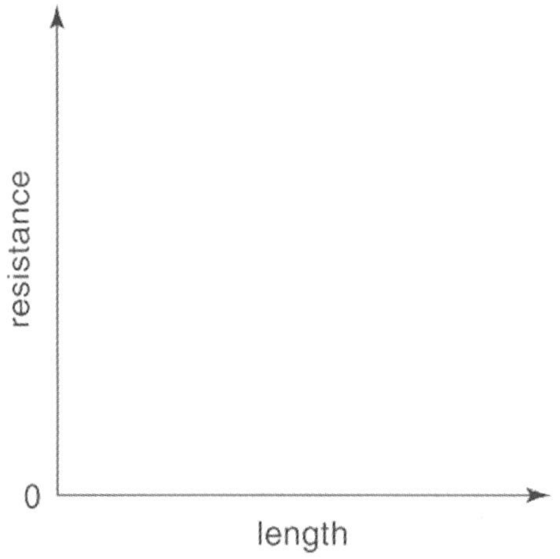

[1 mark]

05.5 **Figure 5.3** shows how the resistance of 1.0 m length of nickel-chromium wire depends on the wire's **thickness**.

Figure 5.3

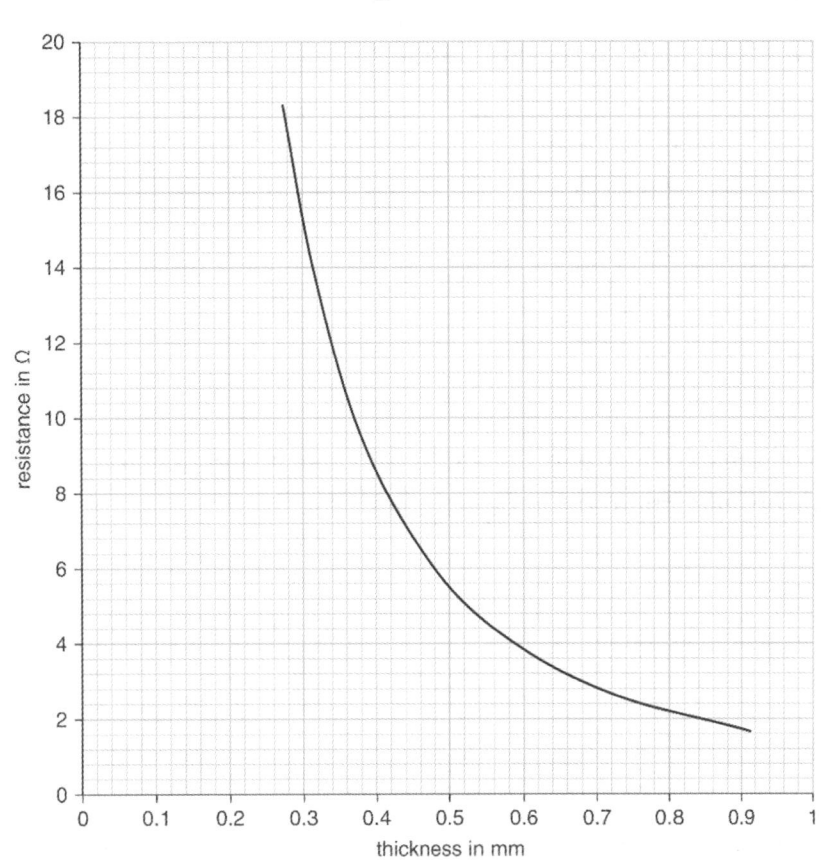

The student suggests that if the thickness of the wire is doubled, its resistance is halved.

Determine whether or not the student's suggestion is correct.

Use data from **Figure 5.3**.

[2 marks]

06

06.1 In some faulty electric toasters, the live wire makes contact with the toaster's metal casing.

Explain why this fault could be very dangerous.

[2 marks]

06.2 Describe how the earth wire in the cable provides protection for anyone attempting to use a toaster with the fault in question **06.1**.

[2 marks]

06.3 An electric power washer is connected to the mains by a long extension cable.

The resistance of the cable is 5.0 Ω

The current in the cable is 8.0 A

Calculate the power wasted as heat in the cable when the power washer is being used.

Give the correct unit with your answer.

Power wasted = ...

Unit: ....

[3 marks]

07

07.1 Compare the advantages and disadvantages of **wind turbines** and **nuclear power** stations for generating electricity in the UK.

Include comparisons of their **reliability** and their **environmental effects**.

[6 marks]

08

08.1 A cyclist cycles uphill at a **steady speed**.

Describe the transfer of energy that is taking place.

....

....

....

[3 marks]

08.2 When the cyclist reaches the top of the hill he stops.

The height of the hill is 10 m

The mass of the cyclist and his bicycle is 100 kg

Gravitational field strength is 10 N/kg

Calculate the increase in the gravitational potential energy store of the cyclist and his bicycle.

....

....

Increase in gravitational potential energy store = ... J

[2 marks]

08.3 The cyclist freewheels down the other side of the hill.

As he freewheels down the hill, 50% of his (and his bicycle's) gravitational potential energy is dissipated to the surroundings.

Calculate the kinetic energy store of the cyclist and his bicycle at the bottom of the hill.

Kinetic energy store = ... J

[2 marks]

08.4 Calculate the speed of the cyclist as he reaches the bottom of the hill.

Give your answer to a suitable number of significant figures.

Speed = ... m/s

[3 marks]

END OF QUESTIONS

Physics Equation Sheet

Equation Number	Word Equation	Symbol Equation
1	(final velocity)² − (initial velocity)² = 2 × acceleration × distance	$v^2 - u^2 = 2as$
2	elastic potential energy = 0.5 × spring constant × (extension)²	$E_e = \frac{1}{2}ke^2$
3	change in thermal energy = mass × specific heat capacity × temperature change	$\Delta E = mc\Delta\theta$
4	period = $\frac{1}{\text{frequency}}$	
5	force on a conductor (at right angles to a magnetic field) = magnetic flux density × current × length	$F = BIl$
6	thermal energy for a change of state = mass × specific latent heat	$E = mL$
7	potential difference across primary coil × current in primary coil = potential difference across secondary coil × current in secondary coil	$V_p I_p = V_s I_s$

Collins

AQA
GCSE
Combined Science: Trilogy H
SET A – Physics: Paper 6 Higher Tier

Author: Lynn Pharoah

Materials

Time allowed: 1 hour 15 minutes

For this paper you must have:
- a ruler
- a calculator
- the Physics Equation Sheet (found at the end of the paper).

Instructions

- Answer **all** questions in the spaces provided.
- Do all rough work in this book. Cross through any work you do not want to be marked.

Information

- There are 70 marks available on this paper.
- The marks for questions are shown in brackets.
- You are expected to use a calculator where appropriate.
- You are reminded of the need for good English and clear presentation in your answers.
- When answering questions 02.1 and 07.2 you need to make sure that your answer:
  – is clear, logical, sensibly structured
  – fully meets the requirements of the question
  – shows that each separate point or step supports the overall answer.

Advice

- In all calculations, show clearly how you work out your answer.

Name: _____

01 Figure 1.1 shows the electromagnetic spectrum.

Figure 1.1

| Gamma rays | X-rays | Ultraviolet | Visible light | Infrared | Microwaves | Radio waves |

01.1 Which group in the electromagnetic spectrum **has the longest wavelength**?

[1 mark]

01.2 Which group in the electromagnetic spectrum **can be detected with the human eye**?

[1 mark]

01.3 Which **two** groups in the electromagnetic spectrum are used to cook food?

.... and ....

[2 marks]

01.4 An X-ray image showing a broken bone is produced by directing a beam of X-rays at the injured part of the body.

The X-rays that pass through the body are detected electronically to form an image.

Explain why X-rays are suitable for investigating a possible broken bone.

[2 marks]

01.5 Satellites used for television broadcasting orbit the Earth at a height well above the Earth's atmosphere.

Communication between the Earth and the satellite uses microwaves with a wavelength of about 0.1 m

Figure 1.2 provides information about the absorption of electromagnetic radiation by the Earth's atmosphere.

Figure 1.2

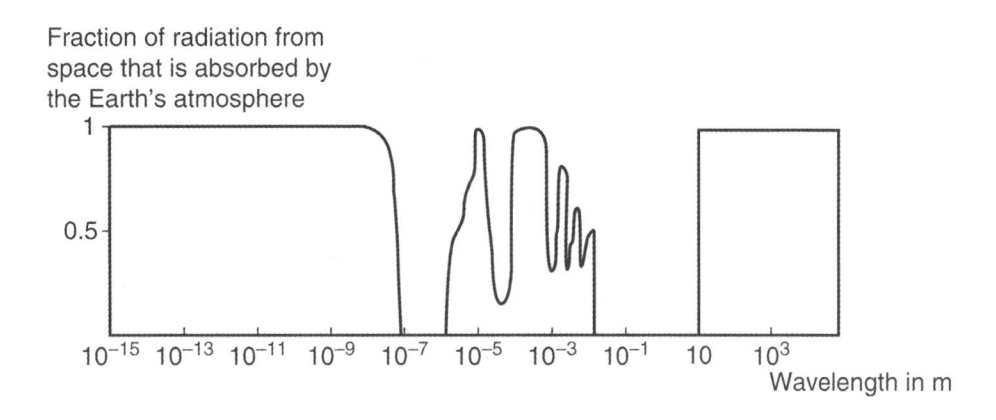

Explain why microwaves of wavelength 0.1 m are used for satellite TV, rather than radio waves which have wavelengths of about 100 m

Use the information in **Figure 1.2**.

[2 marks]

01.6 What can be concluded from **Figure 1.2** about the transmission of **gamma radiation** through the Earth's atmosphere?

[1 mark]

Turn over >

02

02.1 Two of the forces acting on a cyclist in motion are **air resistance** and **rolling resistance** between the bicycle's tyres and the road.

Figure 2.1 shows how these two forces vary with the cyclist's speed.

Figure 2.1

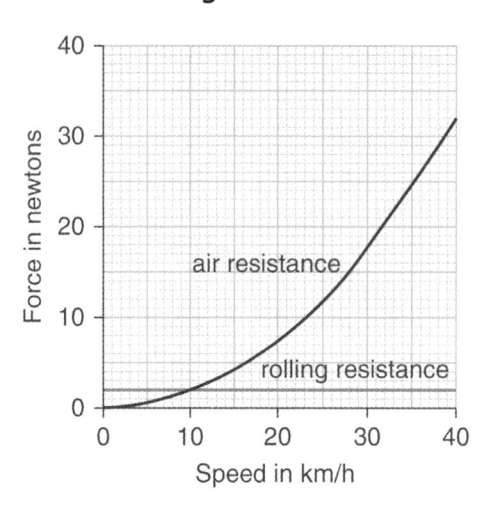

Explain how air resistance and rolling resistance affect the motion of a cyclist.

Use **Figure 2.1** to describe how these resistive forces vary with the speed of the cyclist.

[4 marks]

02.2 Determine the **total** resistive force acting on the cyclist when he is travelling at a steady speed of 20 km/hour.

Use data from **Figure 2.1**.

Total resistive force = _____ N

[3 marks]

02.3 The horizontal forces acting on the cyclist are:

- the total resistive force and
- the driving force he generates by pedalling.

Which of the statements below is correct when he is travelling at **a steady speed**?

Tick **one** box.

The driving force is **larger than** the total resistive force. ☐

The driving force is **equal to** the total resistive force. ☐

The driving force is **smaller than** the total resistive force. ☐

[1 mark]

02.4 Write down the equation which links work done, force and distance.

[1 mark]

02.5 The cyclist maintains a steady speed of 20 km/h over a distance of 200 m.

Calculate the work done by the cyclist.

Give a suitable unit with your answer.

Work done = _____

Unit: _____

[3 marks]

Turn over >

03

03.1 **Figure 3.1** shows the magnetic field between two magnets.

The faces of the magnets are the poles and they are positioned with opposite poles facing.

Figure 3.1

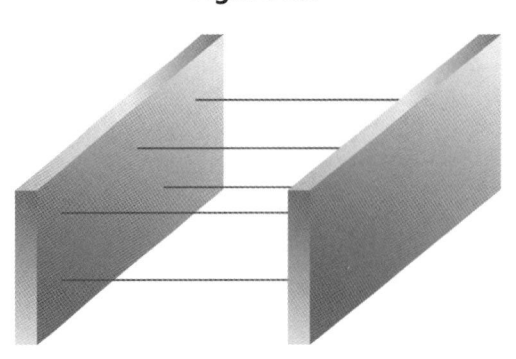

Which **two** words correctly describe the force between magnets?

Tick **two** boxes.

Attractive ☐

Contact ☐

Repulsive ☐

Non-contact ☐

[2 marks]

03.2 Add arrows to **two** of the field lines in **Figure 3.1** to show the direction of the magnetic field between the magnets.

[1 mark]

03.3 A wire carrying an electric current is placed between the poles of the two magnets (**Figure 3.2**).

Figure 3.2

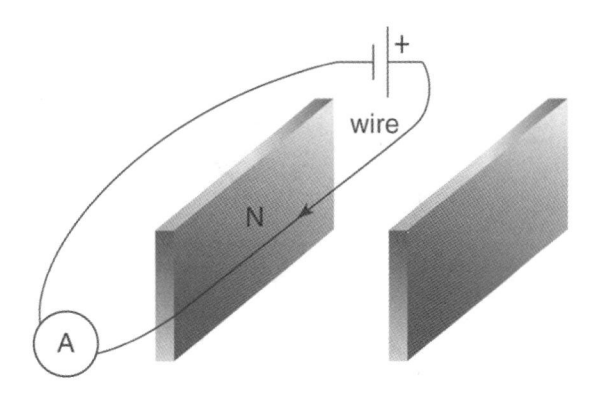

The arrow on the wire shows the direction of the current.

Use Fleming's left-hand rule to determine the direction of the magnetic force on the wire.

The direction of the force on the wire is _____

[1 mark]

Question 3 continues on the next page

03.4 **Figure 3.3** is a sketch graph to show how the force on the wire changes with the size of the current in the wire.

Figure 3.3

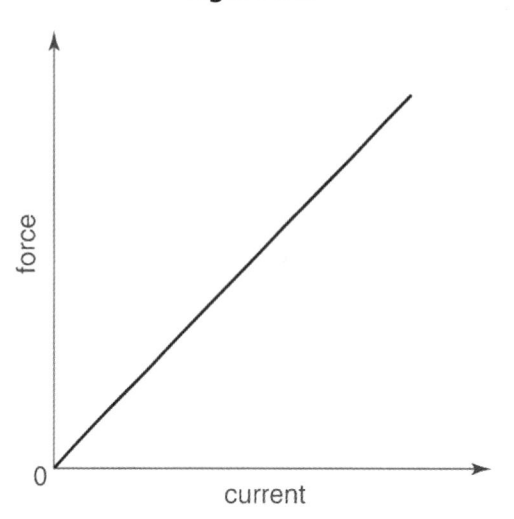

Write a conclusion about the relationship between the force and the current.

....

....

[1 mark]

03.5 The current in the wire in **Figure 3.2** is 3.2 A

The magnetic flux density of the magnetic field is 0.20 T

The length of wire within the field is 0.050 m

Calculate the size of the force on the wire.

Select the correct equation from the Physics Equation Sheet.

....

....

Force = ... N

[2 marks]

04

04.1 Explain the difference between the terms **distance** and **displacement**.

....

....

[1 mark]

04.2 In a 400 m race, an athlete completes one circuit of an athletics track in 50 s

Determine the athlete's average speed.

....

....

Average speed = ... m/s

[2 marks]

04.3 Give the athlete's total displacement.

....

....

Total displacement = ... m

[1 mark]

04.4 How long would it take a typical cyclist to cover a distance of 600 m?

Tick **one** box.

Approximately 10 s ☐

Approximately 100 s ☐

Approximately 1000 s ☐

[1 mark]

Question 4 continues on the next page

04.5 A car is driven along a straight road.

Figure 4.1 shows the car's velocity–time graph during the first 20 s of its motion.

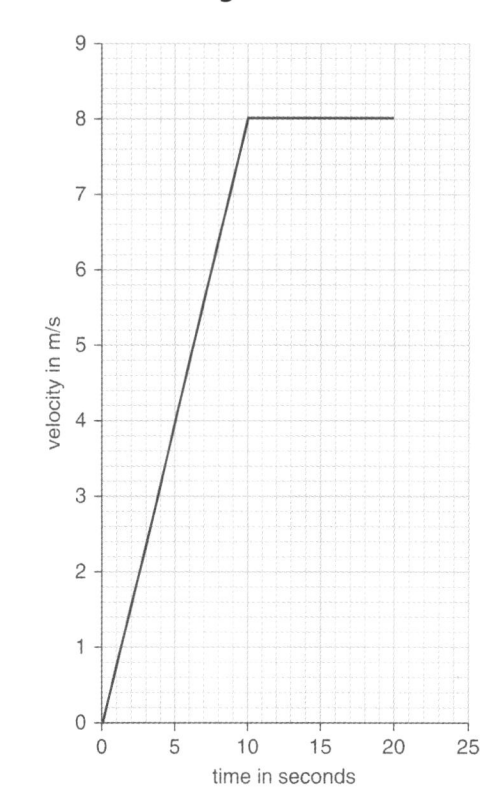

Figure 4.1

Describe the car's motion over the 20 s shown in **Figure 4.1**.

[2 marks]

04.6 Determine the **total distance travelled** by the car during the 20 s shown in **Figure 4.1**.

Distance travelled = _____ m

[3 marks]

05

05.1 **Figure 5.1** shows a plan view of a water tank used to investigate the behaviour of waves.

Figure 5.1

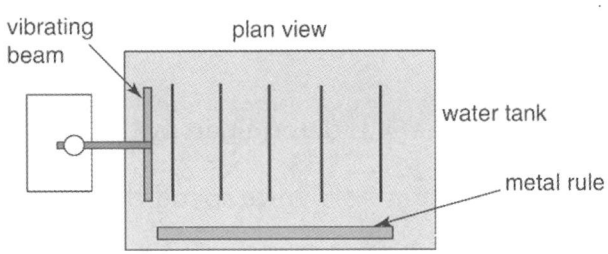

Describe how the waves on the water in the tank are produced.

[1 mark]

05.2 Explain what determines their **frequency**.

[2 marks]

05.3 Explain what is meant by the **wavelength** of a wave.

[1 mark]

Question 5 continues on the next page

05.4 Explain what is meant by the **amplitude** of a wave.

[1 mark]

05.5 The speed of the waves in the tank is determined by the depth of the water.

Figure 5.2 shows a sketch graph of how wave speed changes with water depth.

Figure 5.2

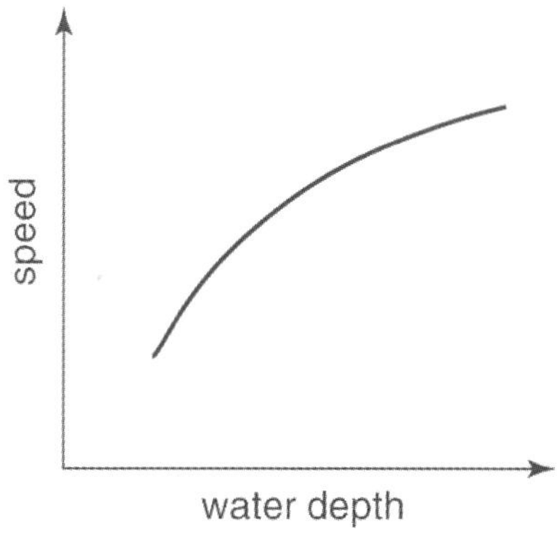

Give **two** conclusions from the graph.

1.

2.

[2 marks]

05.6 Write down the equation linking wave speed, frequency and wavelength.

[1 mark]

05.7 Explain how changing the frequency of the waves in the water tank affects the wavelength. Assume that the speed of the waves is constant.

Your explanation should refer to the wave equation from question **05.5**.

[2 marks]

06

06.1 There is a single point on an object where its weight may be considered to act.

Name this point.

[1 mark]

06.2 Write down the word that is used for the tendency of an object to resist any change to its state of motion.

[1 mark]

06.3 Write down **Newton's first law**, with reference to a **moving** object.

[2 marks]

06.4 **Figure 6.1** is an incomplete free body force diagram of a rocket a few seconds after it has been launched.

There are three forces acting on the rocket: **thrust** from the engine, **air resistance** and **weight**.

Only two of these are shown in **Figure 6.1**.

Figure 6.1

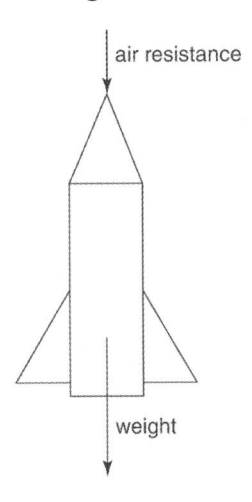

The thrust from the rocket's engine acts vertically upwards.

The rocket is accelerating upwards.

Explain how the sizes of the three forces are related.

....

....

....

....

[2 marks]

Question 6 continues on the next page

06.5 The rocket reaches a height where the effect of air resistance can be ignored.

The engine now produces a **horizontal** thrust of 10 MN, to change the direction of the rocket.

The weight of the rocket is 8 MN

Draw a vector diagram, to scale, of the forces acting on the rocket. Use the grid in **Figure 6.2**.

Using your diagram, determine the direction of the **resultant force** on the rocket.

Show this direction on the diagram.

Figure 6.2

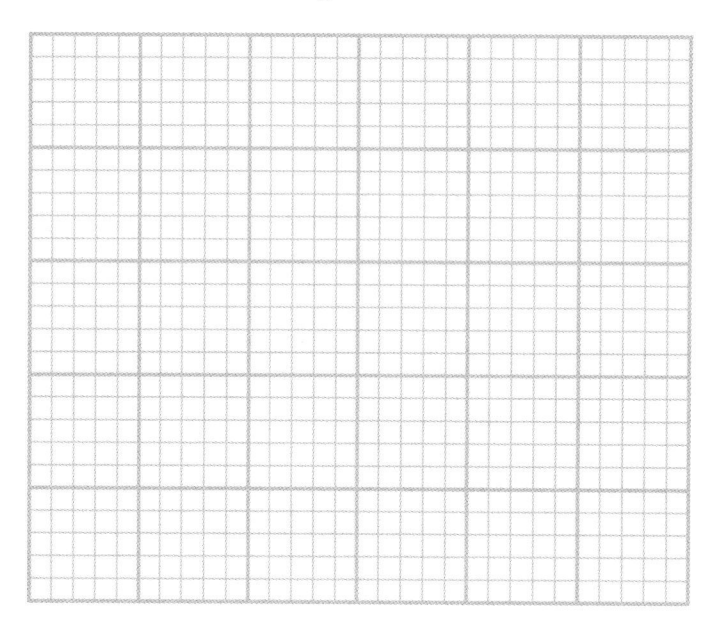

[3 marks]

07 The extension of a spring depends on the stretching force and the spring constant.

A student is asked to use the apparatus in **Figure 7.1** to investigate how the extension of a spring depends on the stretching force applied.

Figure 7.1

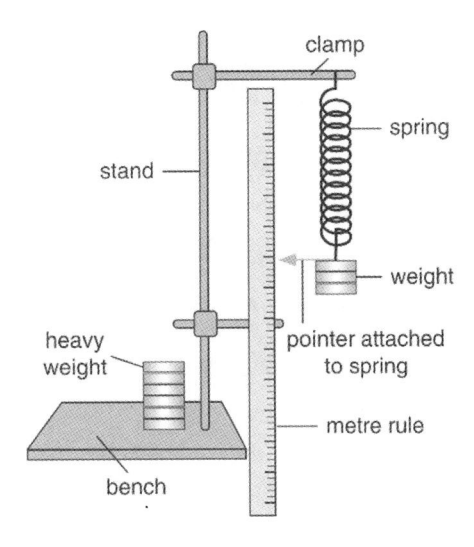

07.1 Identify the variables in this investigation.

Independent variable: _____

Dependent variable: _____

Control variable: _____

[3 marks]

Question 7 continues on the next page

07.2 Write a set of instructions for the student, using the apparatus in **Figure 7.1**.

Include methods of keeping errors to a minimum.

Include a description of how the resulting data should be analysed.

[6 marks]

07.3 A 10 N force stretches a spring by 0.20 m

The spring is **not** inelastically deformed.

Calculate the elastic potential energy stored in the spring.

Elastic potential energy = ... J

[4 marks]

END OF QUESTIONS

Physics Equation Sheet

Equation Number	Word Equation	Symbol Equation
1	(final velocity)² − (initial velocity)² = 2 × acceleration × distance	$v^2 - u^2 = 2\,a\,s$
2	elastic potential energy = 0.5 × spring constant × (extension)²	$E_e = \dfrac{1}{2} k e^2$
3	change in thermal energy = mass × specific heat capacity × temperature change	$\Delta E = m\,c\,\Delta\theta$
4	period = $\dfrac{1}{\text{frequency}}$	
5	force on a conductor (at right angles to a magnetic field) = magnetic flux density × current × length	$F = B\,I\,l$
6	thermal energy for a change of state = mass × specific latent heat	$E = m\,L$
7	potential difference across primary coil × current in primary coil = potential difference across secondary coil × current in secondary coil	$V_p I_p = V_s I_s$

Collins

AQA
GCSE
Combined Science: Trilogy H
SET A – Higher Tier

Authors: Mike Smith, Sunetra Berry and Lynn Pharoah

Answers

Acknowledgements

The author and publisher are grateful to the copyright holders for permission to use quoted materials and images.

All images are © HarperCollinsPublishers and Shutterstock.com

Every effort has been made to trace copyright holders and obtain their permission for the use of copyright material. The author and publisher will gladly receive information enabling them to rectify any error or omission in subsequent editions. All facts are correct at time of going to press.

Published by Collins
An imprint of HarperCollinsPublishers
1 London Bridge Street
London SE1 9GF

© HarperCollinsPublishers Limited 2018
ISBN 9780008282769
First published 2018
10 9 8 7 6 5 4 3 2 1

All rights reserved. No part of this publication may be reproduced, stored in a retrieval system, or transmitted, in any form or by any means, electronic, mechanical, photocopying, recording or otherwise, without the prior permission of Collins.

British Library Cataloguing in Publication Data.

A CIP record of this book is available from the British Library.

Commissioning Editor: Rachael Harrison
Project Leader and Management: Natasha Paul and Katie Galloway
Authors: Mike Smith, Sunetra Berry and Lynn Pharoah
Cover Design: Paul Oates
Inside Concept Design: Ian Wrigley
Text Design and Layout: QBS Learning
Production: Lyndsey Rogers
Printed in Martins the Printers Ltd

Biology: Paper 1

Question	Answer(s)	Extra info	Mark(s)	AO/Spec ref.
01.1	Platelets — Forms clots to seal wounds Skin — Acts as a barrier Stomach — Produces acid to kill pathogens Trachea and bronchi — Secretes mucus to trap pathogens all four correct for **3** marks two or three correct for **2** marks one correct for **1** mark		3	AO1 4.2.2.3 4.3.1.6
01.2	advantage: kill bacteria (inside body) disadvantage: do not kill viruses	allow cures bacterial infection allow may lead to antibiotic resistant strains	1 1	AO1 4.3.1.8
01.3	introduce dead/inactive/harmless/part of pathogens stimulates white blood cells to produce antibodies in future if same pathogens re-enter the body, white blood cells produce antibodies very quickly and in large numbers pathogens killed before can spread/cause symptoms		1 1 1 1	AO1 4.3.1.7
02.1	60 mm = 60 000 μm actual size = image size ÷ magnification = 60 000 ÷ 5000 = 12 (μm)	allow 12 with no working shown for **4** marks allow equivalent marking points if conversion to mm is done at the end	1 1 1 1	AO2 4.1.1.5
02.2	resolution is the ability to distinguish between two points magnification is how many times bigger the image is than the object		1 1	AO1 4.1.1.5
02.3	(not an animal) because it contains chloroplasts (not a plant) because it does not have a cell wall / it does not have a (large/permanent) vacuole (not a bacterium) because it has a nucleus / has chloroplasts / does not have a cell wall / does not contain plasmids/does not have a naked loop of DNA		1 1 1	AO2 4.1.1.1 4.1.1.2

Question	Answer(s)	Extra info	Mark(s)	AO/Spec ref.
03.1	**Level 3:** a coherent method is described with relevant detail, which demonstrates a broad understanding of the relevant scientific techniques and procedures. The steps in the method are logically ordered. The method would lead to the collection of valid results		5–6	AO2/ AO3 4.2.2.1
	Level 2: the bulk of a method is described with mostly relevant detail, which demonstrates a reasonable understanding of the relevant techniques and procedures. The method may not be in a completely logical sequence and may be missing some detail		3–4	
	Level 1: discrete relevant points are made which demonstrate some understanding of the relevant scientific techniques and procedures. They may lack a logical structure and would not lead to the production of valid results		1–2	
	No relevant content		0	
	Indicative content • independent variable is the temperature • temperature is varied by using different temperature water baths • dependent variable is time to digest all the starch • control variables include: concentration/amount of starch, pH • repeat readings and calculate means • plot graph of results to work out the optimum temperature that would give the shortest time/fastest rate of reaction			
03.2	protein — amino acids/peptides lipid / fat / oil — glycerol and fatty acids		4	AO1 4.2.2.1
03.3	add Benedict's solution heat / put in a hot water bath if sugar is present there is a colour change from blue to brick red/orange		1 1 1	AO1 4.2.2.1
04.1	lack of leaves / chlorophyll means less photosynthesis so less glucose is made for growth / for making other substances necessary for growth		1 1 1	AO2 4.3.1.4 4.4.1.3

Question	Answer(s)	Extra info	Mark(s)	AO/Spec ref.
04.2	method 1: use fungicides	explanation must be correctly linked to method	1	AO1/AO2
	explanation: these kill fungus / rose black spot		1	4.3.1.4
	method 2: remove/destroy infected leaves	it does **not** matter which is method 1 or method 2	1	
	explanation: so they cannot act as a source of infection		1	
05.1	% change = $\dfrac{\text{change} \times 100}{\text{original mass}}$		1	AO2 4.1.3.2
	= $\dfrac{(29.0 - 24.0) \times 100}{24.0}$	allow 20.8 with no working shown for **3** marks	1	
	= (+) 20.8 (%)	deduct 1 mark for incorrect rounding	1	
05.2	all points correctly plotted **2 marks**	allow +/- half a small square	2	AO2 4.1.3.2
	but three or four points correctly plotted **1 mark**		1	
	smooth line of best fit			
05.3	correct reading from graph of where line crosses horizontal axis	allow +/- half a small square	1	AO3 4.1.3.2
05.4	as one of the control variables		1	AO2
	surface area (:volume ratio) affects rate of diffusion by osmosis/osmosis		1	4.1.3.1 4.1.3.2
05.5	otherwise would include mass of solution in results / otherwise measurements of mass would be too high		1	AO2 4.1.3.2
06.1	fatty material builds up		1	AO1
	inside coronary arteries		1	4.2.2.4
	reducing blood flow through coronary arteries/to heart muscle		1	
	reducing supply of oxygen/ glucose to heart muscle		1	

Question	Answer(s)	Extra info	Mark(s)	AO/Spec ref.
06.2	**Level 3:** a detailed and coherent evaluation is provided which considers arguments on both sides as to whether the graph demonstrates that obesity is a risk factor for Type 2 diabetes, and comes to a conclusion consistent with the reasoning		5–6	AO3 4.2.2.6
	Level 2: an attempt to give arguments on both sides as to whether the graph demonstrates that obesity is a risk factor for Type 2 diabetes. The logic may be inconsistent at times but builds towards a coherent argument.		3–4	
	Level 1: discrete relevant points made. The logic may be unclear and the conclusion, if present, may not be consistent with the reasoning		1–2	
	No relevant content		0	
	Indicative content • there does appear to be a link between body mass and Type 2 diabetes • but this could simply be correlation not causation • need more evidence, e.g. of a causal mechanism • not a perfect correlation • there may be other factors also linked with Type 2 diabetes • although graph shows body mass it does not show obesity • it's only over a 10-year period • we do not know where the data came from • we do not know how many people were involved			
07.1	**W**: the limiting factor is light intensity		1	AO2 4.4.1.2
	explanation: if you increase light intensity the rate of photosynthesis increases		1	
	X: the limiting factor is carbon dioxide concentration		1	
	explanation: if you increase carbon dioxide concentration the rate of photosynthesis increases		1	
	Y: the limiting factor is temperature		1	
	explanation: if you increase temperature the rate of photosynthesis increases		1	

Question	Answer(s)	Extra info	Mark(s)	AO/Spec ref.
07.2	**Level 3:** a detailed and coherent description of the tests that would have to be made and the conclusions that could be drawn depending on the outcomes		5–6	AO2/ AO3 4.4.1.2
	Level 2: an attempt to give a description of the tests that would have to be made and the conclusions that could be drawn depending on the outcomes. The logic may be inconsistent at times but builds towards a coherent argument		3–4	
	Level 1: discrete relevant points made. The logic may be unclear and any conclusions, if present, may not be consistent with the reasoning		1–2	
	No relevant content		0	
	Indicative content • limiting factor could be carbon dioxide concentration, temperature or amount of chlorophyll • raise the temperature (above 25 °C) but leave carbon dioxide concentration (4%) the same • if the rate of photosynthesis increases then the limiting factor at **Z** is temperature • raise carbon dioxide concentration (above 4%) but leave the temperature (25 °C) the same • if the rate of photosynthesis increases then the limiting factor at **Z** is carbon dioxide concentration • if neither raising carbon dioxide concentration nor temperature increase the rate of photosynthesis then the limiting factor is another factor / the amount of chlorophyll			

Biology: Paper 2

Question	Answer(s)	Extra info	Mark(s)	AO/Spec ref.
01.1	0		1	AO2 4.7.2.1
01.2	4		1	AO2 4.7.2.1
01.3	all points correctly plotted **3 marks** **but** at least 10 points correctly plotted **2 marks** **but** at least six points correctly plotted **1 mark** points joined up to make a 'kite'	allow ± half a small square	3 1	AO2 4.7.2.1

Question	Answer(s)	Extra info	Mark(s)	AO/Spec ref.
01.4	**Level 2:** a detailed and coherent argument is given, which explains why species B and D are more common on the path and why species A and C are more common away from the path		3–4	AO3 4.7.1.1 4.7.2.1
	Level 1: discrete relevant points are made, although the arguments may not be clear		1–2	
	No relevant content		0	
	Indicative content • species A and C are tall(er) • species A and C are killed by mowing on the path • species A and C can survive away from the path as they are tall enough to successfully compete for light • species B and D are low-growing • species B and D are not killed by mowing on the path/are missed by the mower • species B and D cannot survive away from the path as they are not tall enough to successfully compete for light			
02.1	**Level 3:** a coherent evaluation is given, with relevant details, which demonstrates an understanding of the principles of investigations and analysis of results		5–6	AO3 4.5.2
	Level 2: an evaluation is given with mostly relevant detail, which demonstrates a reasonable understanding of the relevant principles. The evaluation may not be completely logical and may be missing some detail		3–4	
	Level 1: discrete relevant points are made which demonstrate some understanding of the relevant principles		1–2	
	No relevant content		0	
	Indicative content **Method** • only recording the shortest time for each student is not as representative as taking the mean result for each student • only using the right hand means that some students may not be using their dominant hand • different numbers of girls and boys is taken into account by taking mean results • sample sizes are small			

Question	Answer(s)	Extra info	Mark(s)	AO/Spec ref.
	Conclusion • it is correct that the mean time for the girls is less than for the boys • the results for the boys show more variation than for the girls • if the longest boys' result (0.32) is discounted then boys overall have the shortest reaction time • the conclusion is based on a small sample size • the conclusion should only apply to this way of measuring reaction time			
02.2	receptor = ear effector = hand muscles		1 1	AO2 4.5.2
02.3	electric impulses along neurones/nerve cells		1 1	AO1 4.5.2
02.4	no – no mark pressing the button is a conscious action or pressing the button is not an automatic action		1	AO2 4.5.2
03.1	<table><tr><th>Mitosis</th><th>Meiosis</th></tr><tr><td>✓</td><td>✗</td></tr><tr><td>✓</td><td>✗</td></tr><tr><td>✗</td><td>✓</td></tr><tr><td>✓</td><td>✗</td></tr></table> all correct for **2 marks** 4, 5, 6 or 7 correct for **1 mark**		2	AO1 4.6.1.1 4.6.1.2
03.2	sequence: 2, 3, 1, 4 all correct for **2 marks** 2 or 3 correct for **1 mark**		2	AO2 4.6.1.2
03.3	<table><tr><th></th><th>X</th><th>Y</th></tr><tr><td>X</td><td>XX female</td><td>XY male</td></tr><tr><td>X</td><td>XX female</td><td>XY male</td></tr></table> correct gametes correct offspring genotypes correct identification of female offspring correct probability of 0.5 **or** 50% **or** 1 in 2 **or** ½		1 1 1 1	AO1 4.6.1.6
03.4	0.5 **or** 50% **or** 1 in 2 **or** ½		1	AO2 4.6.1.6
04.1	to stimulate egg maturation/development		1	AO1 4.5.3.5
04.2	in a laboratory/in a dish		1	AO1 4.5.3.5

Question	Answer(s)	Extra info	Mark(s)	AO/Spec ref.
04.3	(reason:) success rates are low to increase chance of success (disadvantage:) multiple pregnancy/birth risk to mother/babies		1 1 1 1	AO1 4.5.3.5
04.4	FSH = X LH = W oestrogen = Z progesterone = Y all correct for **3 marks** 2 or 3 correct for **2 marks** 1 correct for **1 mark**		3	AO1 4.5.3.3
05.1	Dd	allow dD	1	AO2 4.6.1.4
05.2	DD		1	AO2 4.6.1.4
05.3	does not have polydactyly	allow normal	1	AO2 4.6.1.4
05.4	no – no mark parents must both be dd child needs to inherit at least one D to have condition		1 1	AO2 4.6.1.4
05.5	**Level 2:** a detailed and coherent argument is given, which states all possible genotypes for A and C, and fully explains the reasoning leading to the conclusions		3–4	AO2 4.6.1.4
	Level 1: discrete relevant points are made, including some of the possible genotypes, although the reasoning may not be clear		1–2	
	No relevant content		0	
	Indicative content **A:** • A = Dd or DD • A has condition so must have at least one D • but not enough information to tell whether A is Dd or DD **C:** • C = Dd • C has condition so must have at least one D • C has a mother (B) who must be dd, so C must have inherited a d from B **or** • C has a child (F) who must be dd, so C must have passed on a d to F			

©HarperCollinsPublishers 2018 Combined Science Set A - Answers 5

Question	Answer(s)	Extra info	Mark(s)	AO/Spec ref.
05.6	if disorder is caused by a dominant allele then each individual carrying the allele is affected by the disorder or if disorder is caused by a recessive allele then heterozygous individuals can carry and pass on the condition even though they are unaffected		1	AO2 4.6.1.5
06.1	variation in size among wrens/some wrens were larger than others		1	AO2 4.6.2.1 4.6.2.2
	variation in size is affected/controlled by different genes		1	
	larger wrens are more likely to survive/live longer than smaller ones		1	
	larger wrens pass on the genes for being larger/genes for being smaller are not passed on		1	
06.2	breed them together		1	AO2 4.6.2.2
	to produce fertile offspring		1	
06.3	first two names/genus and species name are the same because they are the same species		1	AO3 4.6.4
	the different/third name shows that there is a difference or T. hirtensis is a sub-species, hence the different name		1	
06.4	*Troglodytes hirtensis*		1	AO2 4.6.4
	same genus name because similar, but different species name		1	
07.1	**Level 3:** a coherent description and explanation is given, with relevant details, which demonstrates an understanding of the links between predator and prey populations		5–6	AO3 4.7.2.1
	Level 2: a description and explanation is given with mostly relevant detail, which demonstrates a reasonable understanding of the relevant principles. The argument may not be completely logical and may be missing some detail		3–4	
	Level 1: discrete relevant points are made which demonstrate some understanding of the relevant principles		1–2	
	No relevant content		0	
	Indicative content • snowy owls nest/breed when there are peaks in lemming abundance • snowy owls do not nest when lemming abundance is low • this is because snowy owls need lemmings to feed their young/they could not raise young if there were not enough lemmings to eat			

Question	Answer(s)	Extra info	Mark(s)	AO/Spec ref.
	• lemming abundance falls after the years when snowy owls have nested • this is because so many lemmings have been eaten by the snowy owls and their young • lemming numbers begin to rise in the years after snowy owls have nested • this is because there is less predation as there will be fewer snowy owls • there is not a perfect correlation between snowy owl nesting and lemming abundance • for example, the years with the highest number of nests are not the years with the highest lemming abundance • this may be because snowy owls may be forced to breed on the island as they are less able to breed elsewhere			
07.2	respiration by snowy owls produces CO_2		1	AO1 4.7.2.2
	decay of waste/dead bodies		1	
	respiration by microorganisms (responsible for decay)		1	
07.3	by protecting places with a higher biodiversity more species may be protected		1	AO3 4.7.3.1
	places with a low biodiversity have their own unique species which should also be protected		1	

Chemistry: Paper 3

Question	Answer(s)	Mark(s)	AO/Spec ref.
01.1		1	AO1 5.2.2
01.2	all electrons correctly shown on the reactant side	1	AO2 5.2.1.2
	all electrons correctly shown on the product side	1	
	Mg· + ×F× ×F× ⟶ [Mg]²⁺ [×F×]⁻ [×F×]⁻ (2, 8, 2) (2, 7) (2, 7) (2, 8) (2, 8) (2, 8)		
01.3		1	AO1 5.2.3
01.4	simple covalent form molecules / giant covalent form macromolecules	1	AO3 5.2.2.4 5.2.2.6
	weak intermolecular forces between simple molecules / none in giant covalent	1	
	simple covalent have low melting and boiling points / giant covalent molecules have high melting and boiling points	1	
02.1	Iodine	1	AO1 5.1.2.6

Question	Answer(s)	Mark(s)	AO/Spec ref.
02.2	one pair of electrons shared	1	AO1
	three pairs of electrons remaining on outershell of each atom	1	5.2.1.4
02.3	**Group 1**	2	AO1 5.1.2.5
	any two from:		
	group 1 elements react more vigorously / violently / explosively as you go down the group		
	more fizzing / bubbles as you go down the group		
	example of a reaction (e.g lithium does not burst into flame, but potassium will spontaneously burst into flame with water)		AO1 5.1.2.6.
	Group 7	2	
	any two from:		
	group 7 element higher up the group will displace the halogen in the salt lower down the group		
	example of a reaction (e.g. chlorine will react with potassium iodide to produce grey iodine)		
03.1	iodine is formed at the anode	1	AO2 5.4.3.2
03.2	they are more reactive than carbon / cannot be extracted by reduction with carbon as they are too reactive	1	AO2 5.4.1.3
	it is too expensive to extract them by displacement with a more reactive metal	1	
03.3	**cathode**	2	AO2 5.4.3.4
	any two from:		
	copper ions and hydrogen ions would be attracted		
	copper is less reactive than hydrogen		
	copper metal is formed / brown coating of copper observed		
	anode	2	
	any two from:		
	chloride and hydroxide ions are attracted		
	chlorine is formed / distinctive smell / bleaches litmus paper		
	bubbles of chlorine gas observed		

Question	Answer(s)	Mark(s)	AO/Spec ref.
04.1	products at higher energy than reactants	1	AO1 5.5.1.2
	correct label for activation energy and overall energy change	1	
04.2	activation energy will be lower	1	AO2 5.5.1.2
04.3	total energy of reactants		AO2 5.5.1.3
	(1 × 615) + (4 × 414) + (1 × 368) = 2639 kJ/mol	1	
	total energy of products		
	(1 × 347) + (5 × 414) + (1 × 276) = 2693 kJ/mol	1	
	energy of reaction = total energy of reactants − total energy of products		
	2639 − 2693	1	
	= −54 kJ/mol	1	
	(allow ecf from earlier parts of the calculation)		
05.1	**Level 3:** a coherent method is described with relevant detail, which demonstrates a broad understanding of the relevant scientific techniques, procedures and safety precautions. The steps in the method are logically ordered with the dependent and control variables correctly identified. The method would lead to the production of valid results. For example, students should specify the amounts of solution and metals to be used. They should explain why the use of a polystyrene cup is preferred to a glass beaker. They should be clear that the maximum temperature rise is the one to be recorded. The resolution of the equipment may be referred to	5–6	AO2 5.4.1.2 5.5.1.1

Question	Answer(s)	Mark(s)	AO/Spec ref.
8	**Level 2:** the bulk of a method is described with mostly relevant detail, which demonstrates a reasonable understanding of the relevant scientific techniques, procedures and safety precautions. The method may not be in a completely logical sequence and may be missing some detail. Students may not specify all volumes and mass of metals. They may make general statements, like measure the temperature before and after, without being clear about which temperature at the end should be recorded.	3–4	
	Level 1: simple statements are made which demonstrate some understanding of some of the relevant scientific techniques, procedures and safety precautions. The response may lack a logical structure and would not lead to the production of valid results.	1–2	
	No relevant content	0	
	Indicative content **Method** • record initial temperature of salt solution with thermometer • place salt solution and metal in beaker • allow time for reaction to complete • record final temperature with thermometer • calculate temperature difference **Fair test** • same mass of metal • same surface area of metal • same volume and concentration of metal salt solutions • same materials for reaction beaker **Reliable and precise data** • polystyrene reaction beaker • insulate reaction vessel to prevent heat loss • weighing scales to 2 d.p. to measure mass of metal • metal surface cleaned with sand paper to remove oxide layer • repeat experiments at least three times for reliability and calculate mean temperature difference • use digital thermometer instead of alcohol thermometer if possible		

Question	Answer(s)	Mark(s)	AO/Spec ref.
05.2	both metals are less reactive than magnesium	1	AO3 5.4.1.2
	both metals are more reactive than copper	1	
	metal Y is more reactive than iron	1	
	allow all three marks for a correct reactivity series that summarises the above points (there is more than one series possible)		
05.3	no	1	AO3 5.4.1.2
	there was no change in the reaction with iron(II) sulfate with metal X but there is a greater temperature change compared with metal Y	1	
05.4	test metal X and Y with the same volume and concentration of acid	1	AO3 5.4.1.2
	record temperature change	1	
06.1	$2Fe_2O_3 + 3C \longrightarrow 4Fe + 3CO_2$ correct formulae of reactants correct balancing	1 1	AO2 5.3.1.1
06.2	reduction	1	AO1 5.4.1.3
06.3	number of moles of sulfuric acid = 50 ÷ 1000 × 2 = 0.1 moles	1	AO2 5.3.2.2
	1:1 mole ratio, so 0.1 moles of Fe	1	
	mass of iron = number of moles × M_r = 0.1 × 56 = 5.6 g	1	
06.4	number of moles of Fe = 4 ÷ 56 = 0.0714	1	AO2 5.3.2.2
	allow ecf from previous answer	1	
	1:1 mole ratio, so moles of $FeSO_4$ = 0.0714	1	
	mass of $FeSO_4$ = 0.0714 × 150 = 10.71 g		
06.5	a limiting reactant is the one not in excess as it limits the amount of product made / the reactant that is completely used up / limits the amount of products	1	AO1 5.3.2.4
	Fe is the limiting reactant in this example	1	
06.6	number of moles of iron(II) sulfate = 5 ÷ 150 = 0.033	1	AO2 5.3.2.5
	to make a 0.33 mol/dm³ solution, 50 g must be dissolved in 1000 cm³	1	
	which is the same as 5 g in **100 cm³**	1	
06.7	alloys are mixtures / metals are elements	1	AO1 5.2.2.7
	pure metals have ions in regular layer / alloys have irregular layers	1	
	layers can slide past each other in a metal / cannot slide in an alloy	1	

Question	Answer(s)	Mark(s)	AO/Spec ref.
07.1	ethanoic acid = 60/46 = 1.30 mol/dm³	1	AO2 5.3.2.5
	HCl 12.16/36.5 = 0.33 mol/dm³	1	
07.2	ethanoic acid is a weak acid and releases fewer H⁺ ions	1	AO2 5.4.2.5
	HCl is a strong acid and fully dissociates so more H⁺ ions released, so pH is lower	1	
07.3	oxidised: Mg atoms	1	AO1 5.4.1.4
	reduced: Hydrogen ions, H⁺	1	
07.4	Mg + 2H⁺ ⟶ Mg²⁺ + H₂		AO1/ AO2 5.1.1.1
	correct reactants and products	1	
	correct balancing	1	
07.5	correctly drawn cell with electrolyte labelled, positive anode, negative cathode connected to a cell or battery.	1	AO2 5.4.3.4 5.4.3.5
	cathode reaction = 2H⁺ + 2e⁻ ⟶ H₂	1	
	anode reaction = 4OH⁻ ⟶ O₂ + 2H₂O + 4e⁻	1	
	bubbles formed at both electrodes	1	

Chemistry: Paper 4

Question	Answer(s)	Mark(s)	AO/Spec ref.
01.1	potable water contains low levels of dissolved salts, pure water has no dissolved salts	1	AO1 5.10.1.2 5.8.1.1
	potable water contains low levels of microbes, pure water has no microbes	1	
01.2	none of the sterilising agents remove all the pathogens	1	AO3 5.10.1.2
	any comparative statement with the use of data (e.g. ozone removes 90% of pathogens, whereas chlorine removes less)	1	
01.3	any two from:	2	AO3 5.10.1.2
	same amount of pathogens to start with		
	sterilising agent exposed to pathogens for the same length of time		
	same volume and type of water		

Question	Answer(s)	Mark(s)	AO/Spec ref.
01.4	any four from:	4	AO2 5.10.1.3
	sewage requires sedimentation to produce sewage sludge and effluent		
	sewage sludge needs to be anaerobically digested		
	effluent from sewage sedimentation is treated with aerobic biological treatment		
	sewage treatment takes longer than ground water treatment		
	sewage treatment is more costly		
	sewage sludge may be used and sold as fertiliser		
	sewage sludge may be used to produce methane as a fuel		
02.1	top curve = powder; lower curve = ribbon	1	AO2 AO3 5.6.1.1 5.6.1.2
	appropriate lines of best fit	1	
	omission of anomalous result at 120 s on lower line	1	
02.2	rate = 40 ÷ 120	1	AO2 5.6.1.1
	= 0.333	1	
	cm³/s	1	
02.3	powder has a greater surface area	1	AO2 5.6.1.3
	more frequent collisions between acid particles and the solid	1	
02.4	particles have higher energy, therefore…	1	AO2 5.6.1.3
	…more frequent collisions	1	
	…more successful/effective collisions	1	
03.1	the general formula for an alkane is C_nH_{2n+2}	1	AO1 5.7.1.1
03.2	kerosene, boiling point: 150–250	1	AO2 5.7.1.3
	fuel oil, boiling point: above 350	1	
	petrol, carbon atoms: 5–10	1	
	liquefied petroleum gas, carbon atoms: 1–4	1	
03.3	$C_4H_{10} + 6.5O_2 \rightarrow 4CO_2 + 5H_2O$	allow multiples (2, 11, 8, 10); 1 mark for correct formulae; 1 mark for correct balancing	2	AO2 5.7.1.3

Question	Answer(s)	Mark(s)	AO/Spec ref.
03.4	decane has higher boiling point/ decane is a liquid, methane is a gas at room temperature	1	AO1 5.7.1.3
	decane is less flammable	1	
	decane will undergo incomplete combustion/ decane will burn with a smoky flame	1	
	accept converse statements for methane		
03.5	cracking breaks down **alkane** molecules…	1	AO1 5.7.1.4
	…into smaller molecules	1	
	polymerisation joins **alkene** molecules…	1	
	…to make larger (long chain) molecules	1	
04.1	it is a carefully controlled mixture	1	AO1 5.8.1.2
04.2	$C_{13}H_{18}O_2$ or $C_{12}H_{17}COOH$ (any order of elements)	1	AO2 5.2.1.4
	$(13 \times 12) + (18 \times 1) + (2 \times 16) = 206$ (allow ecf from formula)	1	5.3.1.2
04.3	any three from:	3	AO1 5.8.1.3
	chromatography can separate mixtures		
	has a stationary phase		
	has a mobile phase		
	separation depends on distribution of substances between the phases		
04.4	B has three spots/three ingredients	1	AO3 5.8.1.3
04.5	A and C are pure antibiotics	1	AO3 5.8.1.3
	D contains a mixture of antibiotic A and C	1	
	B contains antibiotic A and C and one other substance	1	
	(allow any other sensible comparable conclusion, e.g. B, C and D all have one ingredient in common, likewise A and B and D, etc.)		
04.6	$R_f = \dfrac{\text{distance moved by substance}}{\text{distance moved by solvent}}$	award 2 marks for correct substitution even if equation is not written down; final value must be correct, but measurements from the diagram may differ	1	AO2 5.8.1.3
	13 mm ÷ 92 mm = 0.1413		1	
	= 0.14		1	

Question	Answer(s)	Mark(s)	AO/Spec ref.
05.1	**Level 3:** a detailed and coherent comparison is given, which demonstrates a broad knowledge and understanding of the key scientific ideas. The response makes logical links between the points raised and uses sufficient examples to support these links. Students will produce a coherent explanation of all the steps needed to make pure copper from both stops, including electrolysis for final purification. Details of chemicals which need to be added are given, e.g. sulfuric acid and use of scrap iron	5–6	AO3 5.10.1.4
	Level 2: a description is given which demonstrates a reasonable knowledge and understanding of the key scientific ideas. Comparisons are made but may not be fully articulated and/or precise. Students may not have all the steps or chemicals. They may refer to displacement without specifying scrap iron. They may omit the use of electrolysis in final purification	3–4	
	Level 1: simple statements are made which demonstrate a basic knowledge of some of the relevant ideas. The response may fail to make comparisons between the points raised	1–2	
	No relevant content	0	
	Phytomining		
	plants are used to absorb copper compounds from low grade soils		
	the plants are harvested and burnt to produce ash (this step is carbon neutral)		
	the ash is added to sulfuric acid to make copper sulfate		
	displacement of scrap iron can be used to remove copper from the copper sulfate produced		
	electrolysis can be used to purify the copper		
	Advantages/disadvantages		
	plants take up a lot of land		
	land could be used to growing crops for food		
	takes many weeks for the copper to be absorbed		
	Bioleaching		
	bacteria are used in large fermenters		
	bacteria break down the copper in the ores to release copper ions in solution; these are called leachate solutions		
	the leachate solutions are treated chemically to make copper sulfate		

Question	Answer(s)	Mark(s)	AO/Spec ref.
	displacement of scrap iron can be used to remove copper from the copper sulfate produced		
	electrolysis can be used to purify the copper		
	Advantages/disadvantages		
	takes up less space as fermenters are used		
	not as time consuming as phytomining		
	more refining is needed to obtain copper sulfate from leachate		
06.1	any two from: cattle farming; use of landfill sites for waste disposal; cultivation of rice paddies; coal mining; fossil fuel mining	2	AO1 5.9.2.2
06.2	radiation is reflected from the Earth as long wavelength radiation	1	AO1 5.9.2.1
	greenhouse gases, such as carbon dioxide and methane, **absorb** the longer wavelength radiation and prevent it from escaping	1	
06.3	in 1984, methane concentration was low (~1630 ppb) and temp was ~15 °C	statements need to link year to methane concentration **and** to temperature	1	AO3 5.9.2.2
	methane concentration rises to ~1775 ppb in 2000 as temp rises steadily to ~15.2 °C (accept other suitable data from graph)		1	
06.4	any three from: very small temperature differences recorded; very hard to measure temperature or methane concentration globally; need results that are peer reviewed; no indication where this data has come from (possible bias); correlation is not the same as causation	3	AO3 5.9.2.2
06.5	bubble the carbon dioxide through limewater	both statements needed for the mark	1	AO1 5.8.2.3
	the limewater turns (from colourless to) cloudy			

Question	Answer(s)	Mark(s)	AO/Spec ref.
07.1	**Level 3**: a detailed and coherent comparison is given, which demonstrates a broad knowledge and understanding of the key scientific ideas. The response makes logical links between the points raised and uses sufficient examples to support these links. Explanation includes the idea that the equilibrium position will shift to oppose any change. Students will identify that the forward reaction is endothermic from the sign, and relate the explanation for temperature accordingly. They will recognise that there is an increase in number of moles of gas from the balanced equation and base their explanation of pressure on this	5–6	AO2 5.6.2.4 5.6.2.6 5.6.2.7
	Level 2: a description is given which demonstrates a reasonable knowledge and understanding of the key scientific ideas. Comparisons are made but may not be fully articulated and/or precise. Students may not identify that the forward reaction is endothermic, or that more moles of gas are produced; however, they may demonstrate an understanding that the equilibrium will shift to oppose any change	3–4	
	Level 1: simple statements are made which demonstrate a basic knowledge of some of the relevant ideas. The response may fail to make comparisons between the points raised	1–2	
	No relevant content	0	
	Decreasing temperature		
	equilibrium will shift to a position to oppose the decrease in temperature		
	decreased temperature will favour the exothermic reaction/reverse reaction		
	equilibrium will shift to the left		
	yield of hydrogen will decrease		
	Increasing pressure		
	more moles of gas on the product/right-hand side		
	equilibrium will shift position to oppose the increase in pressure		
	equilibrium will shift to the side with fewer moles/lower volume		
	equilibrium will shift to the left		
	yield of hydrogen will decrease		

Physics: Paper 5

Question	Answer(s)	Extra info	Mark(s)	AO/Spec ref.
01.1	a random process		1	AO1 6.4.2.1
	in which an unstable nucleus becomes more stable		1	
	by emitting radiation		1	
01.2	beta		1	AO1 6.4.2.1
01.3	(radiation with a high ionising power) produces a large number of ions		1	AO1 6.4.2.1
	along each cm of its path (accept in a given distance)		1	
	alpha radiation has a short range/doesn't pass through solids (accept doesn't pass through paper)		1	
01.4	$^{14}_{6}C \rightarrow\ ^{14}_{7}N\ +\ ^{0}_{-1}e$	1 mark for each correctly substituted number	4	AO2 6.4.2.2
02.1	accept value in the range of 128 to 132 (W/m^2)		1	AO2 6.1.2.1
02.2	accept value in the range of 64 to 68 (W/m^2)		1	AO2 6.1.2.1
02.3	thermal energy is transmitted at a lower rate through a thicker wall either: doubling the wall thickness, halves the rate of transfer of thermal energy or: rate of energy transferred is inversely proportional to thickness	1 mark for a basic conclusion or 2 marks for a clear reference to inverse proportion	2	AO3 6.1.2.1
02.4	the thermal conductivity of brick is higher than that of air	1 mark	2	AO1 6.1.2.1
	fill the gap with a material with a lower thermal conductivity than air	1 mark		
03.1	number of protons = 79		1	AO1 6.4.1.1
	number of neutrons = 118		1	6.4.1.2
03.2	79 electrons		1	AO1 6.4.1.1
	in orbits / energy levels around the nucleus		1	6.4.1.2

Question	Answer(s)	Extra info	Mark(s)	AO/Spec ref.
03.3	**Level 2**: coherent description of the development of the model of the atom based on the data provided	3–4	4	AO3 6.4.1.3 AO1 6.4.1.3
	Level 1: some relevant content	1–2		
	No relevant content	0		
	Indicative content: • most alpha particles pass through without hitting anything • a very small percentage are deflected • most of the atom must be empty space • alpha particles must be hitting a small concentrated mass/positive charge • model needed to be changed to fit new experimental data • nuclear model: positive charge/ most of the mass is in a nucleus			
04.1	energy required to raise the temperature of 1 kg of a material by 1°C		1	AO1 6.1.1.3
04.2	(temperature rise = 80°C) thermal energy thermal energy = 336 000 (J)	1 mark for correct substitution into correct equation 1 mark for evaluation correct answer with no working 2 marks	2	AO2 6.1.1.3
04.3	power = $\dfrac{\text{energy transferred}}{\text{time}}$		1	AO1 6.1.1.4
04.4	$3000 = \dfrac{\text{energy transferred}}{120}$ energy transferred = 3000 × 120 energy transferred = 360 000 (J)	1 mark for substitution 1 mark for rearranging 1 mark for evaluation correct answer with no working 3 marks	3	AO2 6.1.1.4
04.5	either: efficiency = $\dfrac{\text{useful output energy transfer}}{\text{total input energy transfer}}$ or: efficiency = $\dfrac{\text{useful energy output}}{\text{energy input}}$		1	AO1 6.1.2.2

Question	Answer(s)	Extra info	Mark(s)	AO/Spec ref.
04.6	efficiency = $\frac{336000}{360000} \times 100$ efficiency = 93(%)	1 mark for substitution 1 mark for evaluation correct answer with no working 2 marks allow error carried forward from 04.2 or 04.4 additional 1 mark if answer given to 2 sig. figs	2 1	AO2 6.1.2.2
04.7	(thermal) energy is transferred to the body of the kettle		1	AO1 6.1.2.1
05.1	**Level 2:** a detailed and coherent plan covering all the major steps is given. The steps are presented in a logical order that could be followed by another person to obtain valid results	3–4	4	AO2 6.2.1.3
	Level 1: simple statements relating to relevant apparatus or steps are made but may not follow a logical sequence. The plan would not enable another person to obtain valid results	1–2		
	No relevant content		0	
	Indicative content: • measure the length of wire between the crocodile clips • use a metre rule • close the switch • record the reading on the ammeter • record the reading on the voltmeter • divide the voltmeter reading by the ammeter reading to determine the wire's resistance • repeat for different lengths of wire			

Question	Answer(s)	Extra info	Mark(s)	AO/Spec ref.
05.2	random (error)		1	AO3 6.2.1.3
05.3	systematic (error)	accept 'zero error'	1	AO3 6.2.1.3
05.4	deduct 1 mm from each length measurement		1	AO1 6.2.1.3
05.5	straight line with positive gradient passing through origin		1	AO3 6.2.1.3
05.6	suitable data for comparison selected: e.g. at 0.3 mm, resistance = 15.2 Ω (within ± 0.2 Ω) at 0.6 mm, resistance = 3.8 Ω (within ± 0.2 Ω) yes, student is correct because either: (data shows) doubling thickness reduces resistance to $\frac{1}{4}$ or: doubling thickness, reduces resistance by more than half		1 1	
06.1	a person touching the casing could be electrocuted / have a (large) current passing through them.		1 1	AO1 6.2.3.2
06.2	the earth wire is connected to the metal casing so the current passes through the earth wire instead of the person		1 1	AO1 6.2.3.2
06.3	($P = I^2R$) power wasted = $8.0^2 \times 5.0$ = 320 unit: W or watt	1 mark for substitution into correct equation 1 mark for evaluation correct answer with no working 2 marks 1 mark for correct unit	2 1	AO2 6.2.4.1 AO1 6.2.4.1
07.1	**Level 3:** compares wind and nuclear regarding reliability compares wind and nuclear regarding two examples of environmental effects	5–6	6	AO3 6.1.3
	Level 2: describes one advantage and one disadvantage for both wind and nuclear power	3–4		

Question	Answer(s)	Extra info	Mark(s)	AO/Spec ref.
	Level 1: mentions one disadvantage or advantage of either wind or nuclear power	1–2		
	No relevant content		0	
	Indicative content: **Advantages of wind turbines:** • renewable power • usually windy somewhere in the UK • does not cause pollution • no greenhouse gas emissions/ does not contribute as much to climate change **Disadvantages of wind turbines:** • unpredictable • not reliable • possible threat to wildlife • possible noise disturbance • turbines may be considered to have a negative visual impact **Advantages of nuclear power:** • reliable • no greenhouse gas emissions/ does not contribute to climate change **Disadvantages of nuclear power:** • not renewable • creates radioactive waste • workers are exposed to radiation			
08.1	(store of) chemical energy to (store of) gravitational potential energy and (store of) thermal energy		1 1 1	AO1 6.1.1.1
08.2	($\Delta E_p = mgh$) increase in gravitational potential energy store = 100 × 10 × 10 = 10000 (J)	1 mark for substitution into correct equation 1 mark for evaluation correct answer with no working 2 marks	2	AO2 6.1.1.2
08.3	kinetic energy store = 0.5 × 10000 = 5000 (J)	allow error carried forward from 08.2	1 1	AO2 6.1.2.1

Question	Answer(s)	Extra info	Mark(s)	AO/Spec ref.
08.4	($\Delta E_k = \tfrac{1}{2}mv^2$) $5000 = \tfrac{1}{2} \times 100 \times v^2$ $v = \sqrt{\dfrac{2 \times 5000}{100}}$ speed = 10 (m/s)	1 mark for substitution into correct equation 1 mark for rearranging 1 mark for evaluation correct answer with no working 3 marks	3	AO2 6.1.1.2

Physics: Paper 6

Question	Answer(s)	Extra info	Mark(s)	AO/Spec ref.
01.1	radio waves		1	AO1 6.6.2.1
01.2	visible light		1	AO1 6.6.2.1
01.3	infrared microwaves		1 1	AO1 6.6.2.4
01.4	X-rays are absorbed by bony tissue X-rays pass through soft tissue		1 1	AO1 6.6.2.3
01.5	(electromagnetic) radiation of wavelength 0.1 m is not absorbed by the Earth's atmosphere all / 100% of (electromagnetic) radiation of wavelength 100m is absorbed by the Earth's atmosphere		1 1	AO3 6.6.2.2
01.6	gamma radiation is not transmitted (through the atmosphere) or gamma radiation is totally absorbed by the atmosphere		1	AO3 6.6.2.2
02.1	**Level 2:** coherent description of the effect of the two forces on the motion of the cyclist and their relationship with the speed of the cyclist	3–4	4	AO3 6.5.1.2

Question	Answer(s)	Extra info	Mark(s)	AO/Spec ref.
	Level 1: some relevant content but must include a reference to Figure 2.1 to gain the 2 marks	1–2		
	No relevant content	0		
	Indicative content: air resistance and rolling resistance oppose the motion of the cyclist			
	some of the work / effort of the cyclist has to overcome the resistive forces			
	rolling resistance is not affected by the speed of the cyclist			
	air resistance increases with speed			
	the rate at which the air resistance increases with speed gets greater at higher speeds			
02.2	rolling resistance = 2 (N)	1 mark	3	AO2 6.5.1.2
	air resistance = 7.5 (N) (accept 7 to 8)	1 mark		
	total resistive force = 9.5 (N) (accept 9 to 10)	1 mark correct answer with no working 3 marks		
02.3	the driving force is equal to the total resistive force	only one box ticked	1	AO1 6.5.4.2.1
02.4	work done = force × distance (along line of action of force)	accept $W = Fs$ or $W = Fd$	1	AO1 6.5.2
02.5	work done = 9.5 × 200	1 mark for substitution	2	AO2 6.5.2
	work done = 1900 (accept 1800 to 2000)	1 mark for evaluation correct answer with no working 2 marks		
	unit: accept either J or N m		1	AO1 6.5.2
		allow error carried forward from 02.2		
		1 mark for unit		
03.1	attractive non-contact	1 mark each only two boxes ticked	2	AO1 6.7.1.1
03.2	arrows on two lines showing direction from left to right		1	AO1 6.7.1.2
03.3	upwards		1	AO2 6.7.2.2

Question	Answer(s)	Extra info	Mark(s)	AO/Spec ref.
03.4	force is directly proportional to current		1	AO3 6.7.2.2
03.5	($F = BIl$) force = 0.20 × 3.2 × 0.050	1 mark for substitution into correct equation	2	AO2 6.7.2.2
	force = 0.032 (N)	1 mark for evaluation do not accept 0.03 correct answer with no working 2 marks		
04.1	displacement is distance measured in a straight line in a given direction		1	AO1 6.5.4.1.1
	accept displacement is a vector / has direction and distance is a scalar / has no direction			
04.2	400 = average speed × 50	1 mark for substitution and rearranging	2	AO2 6.5.4.1.2
	average speed = $\frac{400}{50}$			
	average speed = 8 (m/s)	1 mark for evaluation correct answer with no working 2 marks		
04.3	displacement = 0 (m)		1	AO1 6.5.4.1.1
04.4	100 s	only one box ticked	1	AO1 6.5.4.1.2
04.5	the car accelerates from rest for the first 10 s		1	AO3 6.5.4.1.5
	then it has a constant velocity / speed of 8 m/s		1	
04.6	attempt to determine the area under graph line	1 mark 1 mark for calculation	3	AO2 6.5.4.1.5
	= (5 × 8) + (8 × 10)			
	distance travelled = 120 (m)	1 mark for evaluation correct answer with no numerical working 2 marks		

Question	Answer(s)	Extra info	Mark(s)	AO/Spec ref.
05.1	the up and down motion of the vibrating beam produces the waves		1	AO1 6.6.1.2
05.2	the frequency of the waves is determined by the number of times the beam goes up and down / vibrates		1	AO1 6.6.1.2
	in one second		1	
05.3	the distance from a point on a wave to the equivalent point on the adjacent wave		1	AO1 6.6.1.2
	allow distance between (adjacent) troughs / distance between (adjacent) peaks			
05.4	maximum displacement (of a point on a wave) from its undisturbed position		1	AO1 6.6.1.2
05.5	wave speed increases as the water gets deeper	1 mark	2	AO3 6.6.1.2
	and: • as depth increases, the wave speed increases by a smaller amount / tends towards a limit or • at greater depths, the rate at which the speed increases with depth is reduced	1 mark for either		
05.6	wave speed = frequency × wavelength	accept $v = f\lambda$	1	AO1 6.6.1.2
05.7	increasing the frequency decreases the wavelength		1	AO1 6.6.1.2
	wave equation shows frequency × wavelength (= speed) does not change / is constant		1	
	accept wavelength and frequency are inversely proportional			
06.1	centre of mass		1	AO1 6.5.1.3
06.2	inertia	accept inertial mass	1	AO1 6.5.4.2.1
06.3	if the resultant force on a (moving) object is zero, the object continues to move at the same speed and in the same direction.	1 mark 1 mark allow 'continues to move with the same velocity' for full marks	2	AO1 6.5.4.2.1

Question	Answer(s)	Extra info	Mark(s)	AO/Spec ref.
06.4	the thrust is greater than the weight and air resistance added together		1	AO2 6.5.1.4
	(or: thrust > weight + thrust)			
	explanation: (for the rocket to be accelerating upwards) there must be a resultant upward force on the rocket		1	
06.5	downwards vertical force (8 MN) and horizontal force (10 MN) drawn as arrows that are the correct length relative to the size of the forces		2	AO2 6.5.1.4
	accept horizontal force to the left or the right, as it is not specified in question		1	
	resultant force constructed correctly and with arrow in correct direction consistent with the two force arrows			
07.1	independent variable: (stretching) force		1	AO3 6.5.3
	dependent variable: extension		1	
	control variable: spring constant (or 'same spring')		1	
07.2	**Level 3:** a coherent account of a method that produces valid data, leading to the plot of a graph of extension versus stretching force at least one reference to minimising errors needed to gain the maximum mark	5–6	6	AO2 6.5.3
	Level 2: a clear method referring specifically to the apparatus shown that would obtain a set of valid data of extension and stretching force at least one reference to minimising errors needed to gain the maximum mark	3–4		
	Level 1: a basic method that would obtain an extension measurement is given but may not refer specifically to the apparatus shown	1–2		
	No relevant content	0		

Question	Answer(s)	Extra info	Mark(s)	AO/Spec ref.
	Indicative content: • with no weight attached, the metre rule reading in line with the pointer is recorded • a (specific) weight is attached to the spring • the metre rule reading in line with the pointer is again recorded • the extension is the difference between the two readings • repeat the procedure with different weights • to minimise errors: 　• view the pointer from the same horizontal level 　• take repeat readings and average • analysis: plot a graph of extension versus stretching force			
07.3	$(F = k\,e)$ $10 = $ spring constant $\times 0.20$ spring constant $= \dfrac{10}{0.20} = 50$ (N/m) $(E_p = 0.5\,k\,e^2)$ elastic potential energy $= \dfrac{1}{2} \times 50 \times (0.20)^2$ elastic potential energy $= 1.0$ (J)	1 mark for substitution into correct equation for spring constant 1 mark for rearranging and evaluation 1 mark for substitution into equation for elastic potential energy from Physics Equation Sheet 1 mark for evaluation allow error carried forward from evaluation of spring constant	4	AO2 6.5.3

©HarperCollins*Publishers* 2018   Combined Science Set A – Answers

BLANK PAGE

BLANK PAGE

Collins

AQA
GCSE
Combined Science: Trilogy H
SET B – Biology: Paper 1 Higher Tier
Author: Kath Skillern

Materials

Time allowed: 1 hour 15 minutes

For this paper you must have:
- a ruler
- a calculator

Instructions

- Answer **all** questions in the spaces provided.
- Do all rough work in this book. Cross through any work you do not want to be marked.

Information

- There are 70 marks available on this paper.
- The marks for questions are shown in brackets.
- You are expected to use a calculator where appropriate.
- You are reminded of the need for good English and clear presentation in your answers.
- When answering questions 06.2 and 08.2 you need to make sure that your answer:
  – is clear, logical, sensibly structured
  – fully meets the requirements of the question
  – shows that each separate point or step supports the overall answer.

Advice

- In all calculations, show clearly how you work out your answer.

Name:

01 Pathogens cause diseases in plants and animals.

Plants and animals are able to defend themselves against attack.

01.1 Name a plant disease that is caused by a virus.

[1 mark]

01.2 Explain how this virus affects the whole plant.

[3 marks]

01.3 Name **two** non-specific defence systems of the human body.

[2 marks]

01.4 Suggest **two** ways in which white blood cells help to defend the human body against pathogens.

[2 marks]

02 The digestive system is a collection of organs that work together to digest and absorb our food.

02.1 What is the name given to biological molecules that break down our food?

Tick **one box.**

Catalysts ☐

Enzymes ☐

Proteins ☐

Substrate ☐

[1 mark]

02.2 Amylase is a carbohydrase which breaks down starch to maltose and glucose.

Tom wanted to investigate the effect of pH on the rate of reaction of amylase.

This is the method used.

1. Gather three solutions:
   - amylase
   - starch solution
   - pH buffer solution.

2. Set up a spotting tile with rows of iodine drops and prepare the stopwatch.

3. Mix the three solutions in a test tube in a particular order and start the stopwatch.

Which is the correct order to put the solutions into the test tube?

_____

[1 mark]

Question 2 continues on the next page

02.3 Explain why it is important that Tom mixed the solutions in the correct order.

....

....

....

[2 marks]

02.4 Tom could have set up a colour control with iodine and water.

Why might a control have helped him?

....

....

....

[2 marks]

03 When using a microscope, live cells can be mounted in a drop of water on a microscope slide.

They are then covered using a transparent coverslip.

03.1 Give **two** reasons for using a coverslip when looking at a slide under the microscope.

[2 marks]

03.2 When using a microscope, there is a difference between the field of view of a low-power lens and the field of view of a high-power lens.

Explain what causes this difference.

[2 marks]

Question 3 continues on the next page

03.3 A micrograph is a photograph taken using a microscope.

Figure 3.1 shows a low-power micrograph of a plant root.

Figure 3.1

Draw a diagram of the plant root.

Label the meristem on your diagram.

Draw a scale bar on your diagram, with units.

[4 marks]

04 An estimated 42% of cancer cases each year in the UK are linked to lifestyle choices.

Look at **Figure 4.1**.

Figure 4.1

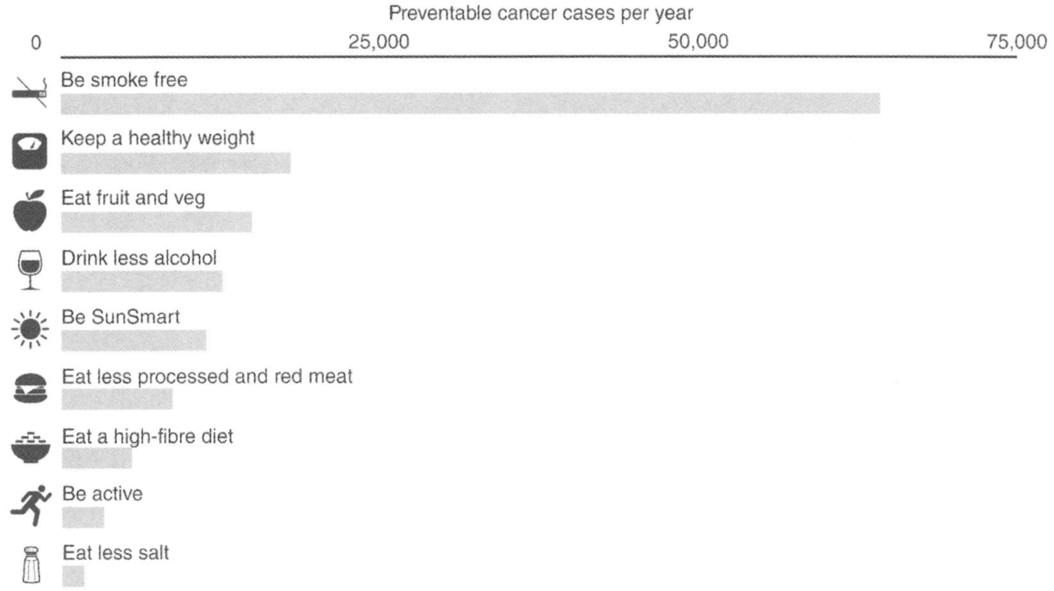

04.1 How many preventable cancers were related to smoking?

[1 mark]

04.2 Compare the numbers of preventable cancers related to **being active** with those related to drinking less alcohol.

[2 marks]

04.3 The most common types of cancers are different for men compared with women.

Suggest reasons for this.

[3 marks]
Turn over >

05 A vaccination introduces a small quantity of dead pathogen into the body to protect us from disease.

A new vaccination has been developed against the pathogen Lumpius.

The Lumpius vaccine is being tested by a pharmaceutical company, which has recruited 10,000 volunteers.

Figure 5.1 shows the body's response to the vaccination and, later, to infection by Lumpius.

Figure 5.1

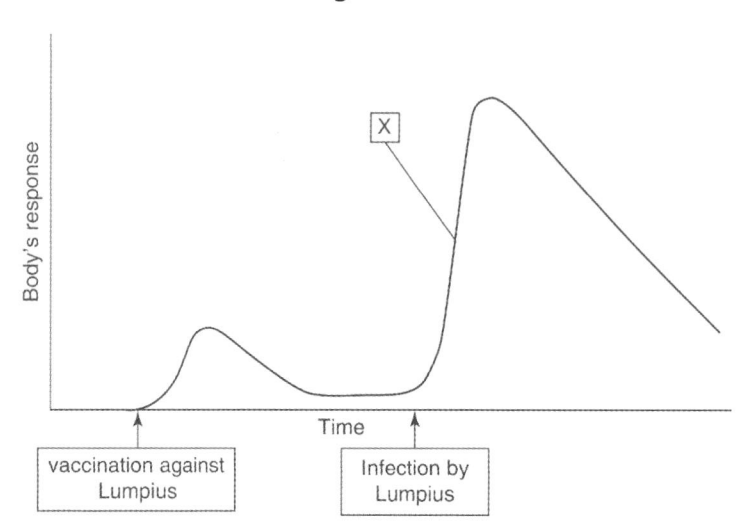

05.1 Explain what is happening at X.

[3 marks]

05.2 Use **Figure 5.1** to draw conclusions about:

- how effective the vaccine is
- the dose of the vaccine.

[4 marks]

05.3 At what stage of development is the vaccine?

Give a reason for your answer.

[2 marks]

06 Jane has set up some apparatus to investigate the rate of photosynthesis in an aquatic plant.

06.1 Look at **Figure 6.1**.

Figure 6.1

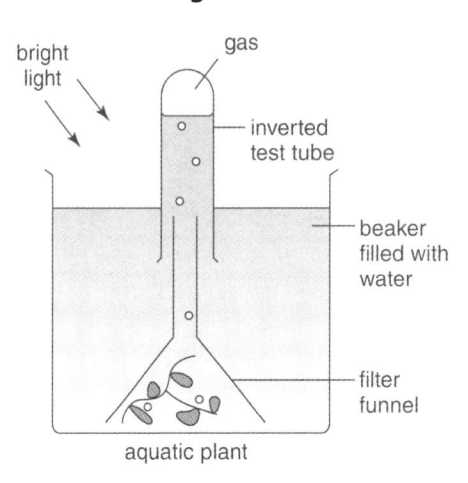

What is the name of the gas collecting in the test tube?

[1 mark]

06.2 Explain how Jane should carry out her investigation.

[6 marks]

06.3 Jane wants a pond in her garden to keep fish.

Explain why she should dig her pond in a sunny part of the garden.

[2 marks]

Question 6 continues on the next page

06.4 **Figure 6.2** shows Jane's results.

Figure 6.2

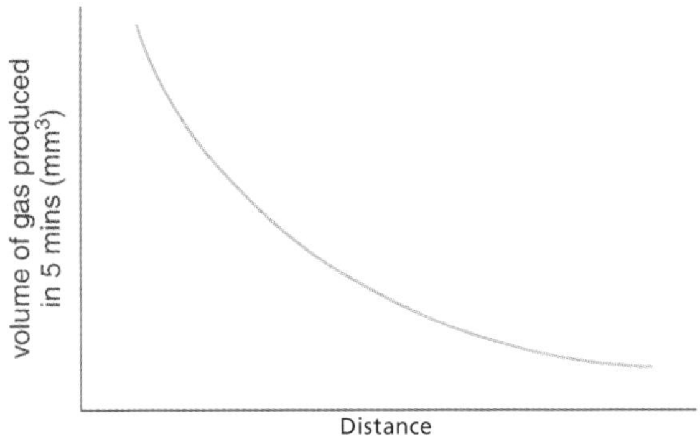
volume of gas produced in 5 mins (mm^3)
Distance

Explain the shape of the graph in **Figure 6.2.**

....

....

....

....

[3 marks]

07 During long periods of vigorous activity, insufficient oxygen is supplied to the muscles and anaerobic respiration takes place.

An oxygen debt is created by a build-up of lactic acid.

07.1 **Figure 7.1** shows oxygen consumption over time.

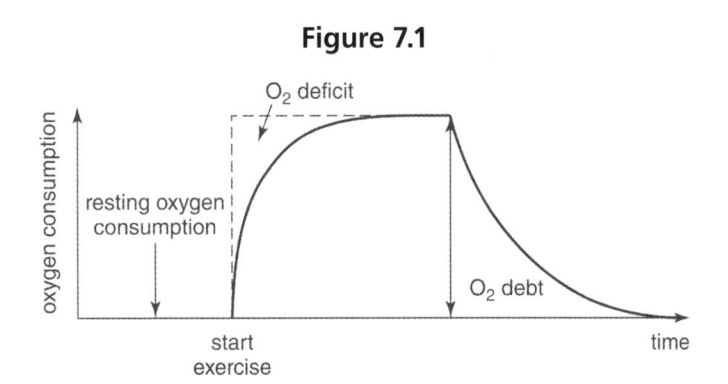

Figure 7.1

Add the following labels to the graph:

    A. Excess post-exercise oxygen consumption

    B. Steady-state oxygen consumption

    C. Oxygen requirement

    D. Finish exercise

    E. Recovery

[5 marks]

Question 7 continues on the next page

07.2 Describe what happens after exercising to the lactic acid that has built up in the muscles.

....

....

....

[2 marks]

07.3 Describe what is meant by 'oxygen debt', in terms of the amount of oxygen required by the body.

....

....

....

[2 marks]

08 Metabolism is the sum of all the reactions in a cell or organism.

The energy transferred supplies all the energy needed for living processes.

One of these processes is respiration.

08.1 Energy from respiration is used in active transport.

What is active transport?

[2 marks]

08.2 Describe other processes of metabolism.

[6 marks]

09 **Figure 9.1** shows worldwide wheat production from 1951 to 1995.

Figure 9.1

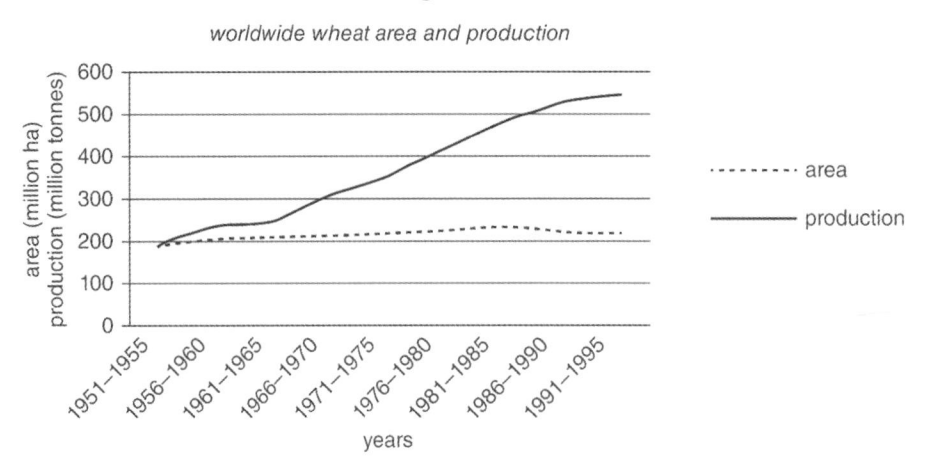

worldwide wheat area and production

09.1 Use **Figure 9.1** to describe the worldwide wheat production in relation to the area of land used to cultivate wheat.

[2 marks]

09.2 Between 1961 and 1965 the wheat production was about 250 million tonnes.

How long did it take for the wheat production to double?

[1 mark]

09.3 Suggest **one** factor that would help to explain these observations.

[1 mark]

END OF QUESTIONS

Collins

AQA
GCSE
Combined Science: Trilogy H
SET B – Biology: Paper 2 Higher Tier

Author: Kath Skillern

Materials

Time allowed: 1 hour 15 minutes

For this paper you must have:
- a ruler
- a calculator

Instructions

- Answer **all** questions in the spaces provided.
- Do all rough work in this book. Cross through any work you do not want to be marked.

Information

- There are 70 marks available on this paper.
- The marks for questions are shown in brackets.
- You are expected to use a calculator where appropriate.
- You are reminded of the need for good English and clear presentation in your answers.
- When answering questions 03.3, 04.2 and 06.1 you need to make sure that your answer:
  – is clear, logical, sensibly structured
  – fully meets the requirements of the question
  – shows that each separate point or step supports the overall answer.

Advice

- In all calculations, show clearly how you work out your answer.

Name: _____

01 An ecosystem is the interaction of a community of living organisms with the non-living parts of their environment.

   01.1 How is the **non-living** part of the environment described?

   Tick **one** box.

   Abiotic ☐

   Biotic ☐

   Dead ☐

   Habitat ☐

   [1 mark]

   01.2 Name **two** resources that animals compete for.

   ....

   ....

   [2 marks]

   01.3 Within a community each species depends on other species to help it survive.

   If one species is removed it can affect the whole community.

   How is this described?

   ....

   [1 mark]

   01.4 What is a **stable community**?

   ....

   ....

   ....

   [2 marks]

01.5 **Figure 1.1** shows a simple food web from a grassland community.

Figure 1.1

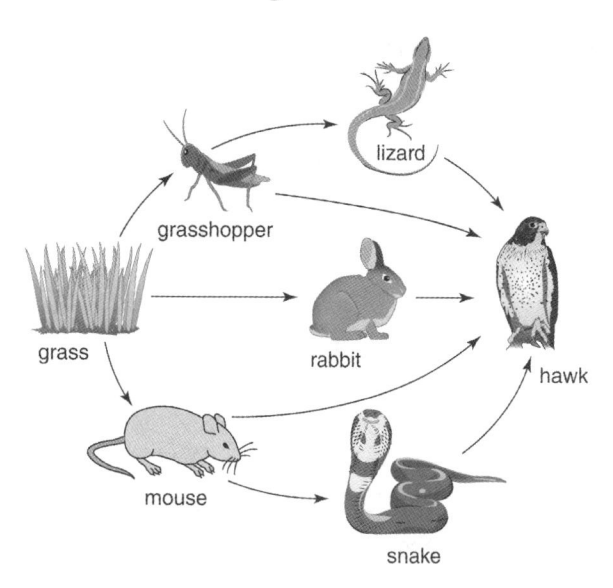

Choose a food chain which includes **four** organisms.

Write down:

- the name of each organism
- the name given to the level of each organism within the ecosystem.

The first level has been started for you.

1. __Grass_____ is a __primary_____ producer

2. _____ is a _____ _____

3. _____ is a _____ _____

4. _____ is a _____ _____

[4 marks]

02 Evolutionary trees are used by scientists to show how organisms are related.

Figure 2.1 shows an evolutionary tree.

The numbers on the branches of the evolutionary tree are the number of 'million years ago'.

Figure 2.1

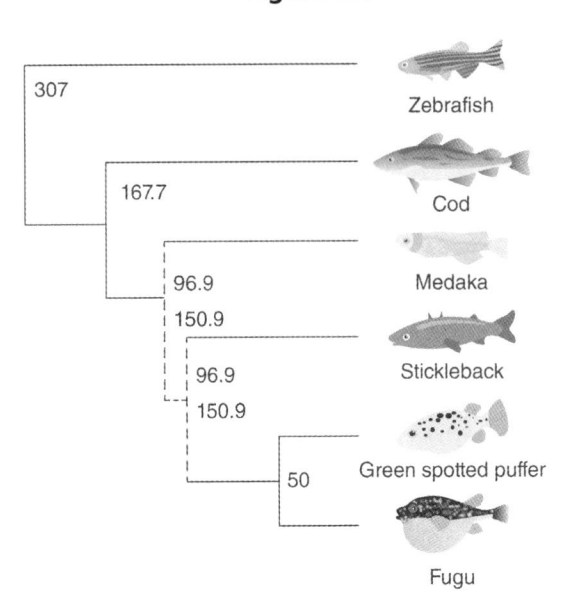

02.1 Which two fishes are most **closely** related?

Tick **two** boxes.

Cod ☐

Fugu ☐

Green spotted puffer ☐

Medaka ☐

Stickleback ☐

Zebrafish ☐

[1 mark]

02.2 How long ago did the cod split from medaka and stickleback?

[1 mark]

02.3 Suggest why there is only a **dotted** line between medaka and stickleback.

[1 mark]

02.4 Name **one** type of evidence that helps scientists construct evolutionary trees.

[1 mark]

02.5 Describe the **key ideas** in the theory of evolution by natural selection.

[4 marks]

02.6 Each gene may have different forms called alleles.

When is a recessive allele expressed?

[1 mark]

Turn over >

03 Type 2 diabetes is a serious condition.

In Type 2 diabetes the body's cells no longer respond as effectively to control glucose concentration in the blood.

Look at **Table 3.1**.

Table 3.1

Year	Mean body mass (kg)	Proportion of the population who have Type 2 diabetes (%)
1990	72.5	4.9
1991	73.0	5.0
1992	73.7	5.4
1993	74.0	4.7
1994	74.6	5.3
1995	75.0	5.5
1996	74.8	5.4
1997	75.3	6.2
1998	76.0	6.5
1999	76.6	6.9
2000	77.2	7.3

03.1 Use the data in **Table 3.1** to plot a graph to show the effect of body mass on percentage of the population who have Type 2 diabetes.

Make sure to:

- choose an appropriate scale
- label both the axes
- plot all points to show the pattern of results.

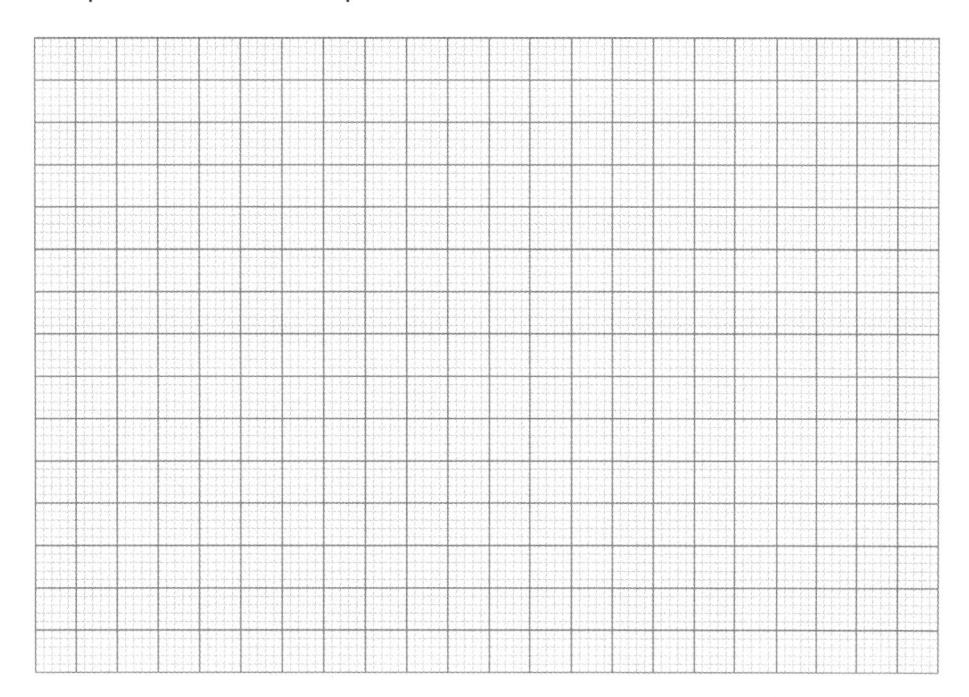

[4 marks]

03.2 Describe the relationship between the mean body mass of the population and the percentage of people who have **Type 2 diabetes.**

[1 mark]

Question 3 continues on the next page

03.3 If one person has Type 2 diabetes, and another person does not:

- explain how the negative feedback system in their bodies controls high levels of blood glucose concentration
- describe the differences in the blood glucose concentration of the two people after they have both eaten a full breakfast.

[6 marks]

04 The human body reacts to the changes by coordinating a **nervous** response or a **hormonal** response.

04.1 Describe the ways in which the body's **hormonal response** is different from the **nervous response**.

[2 marks]

04.2 Describe the reflex reaction in a human when a hot flame is detected on the palm of the hand.

[6 marks]

Question 4 continues on the next page

04.3 In a scientific study, reaction times were investigated after four volunteers had drunk alcohol.

A small can of beer contains about one unit of alcohol.

The results are shown in **Table 4.1**.

Table 4.1

Volunteer	Reaction time in milliseconds (ms)				
Units of alcohol	0.5	1.5	3.0	4.5	6.0
A	34	45	59	71	85
B	35	47	62	75	87
C	32	46	64	72	83
D	30	42	59	70	
Mean	33	45	61	72	84

Calculate the reaction time of volunteer D after 6.0 units of alcohol.

Reaction time of volunteer D after 6.0 units of alcohol = _____

[3 marks]

04.4 What do these results suggest about the effect of drinking alcohol on reaction times?

[1 mark]

05 Figure 5.1 shows five closely related species of fish, with their diets and habitats.

Figure 5.1

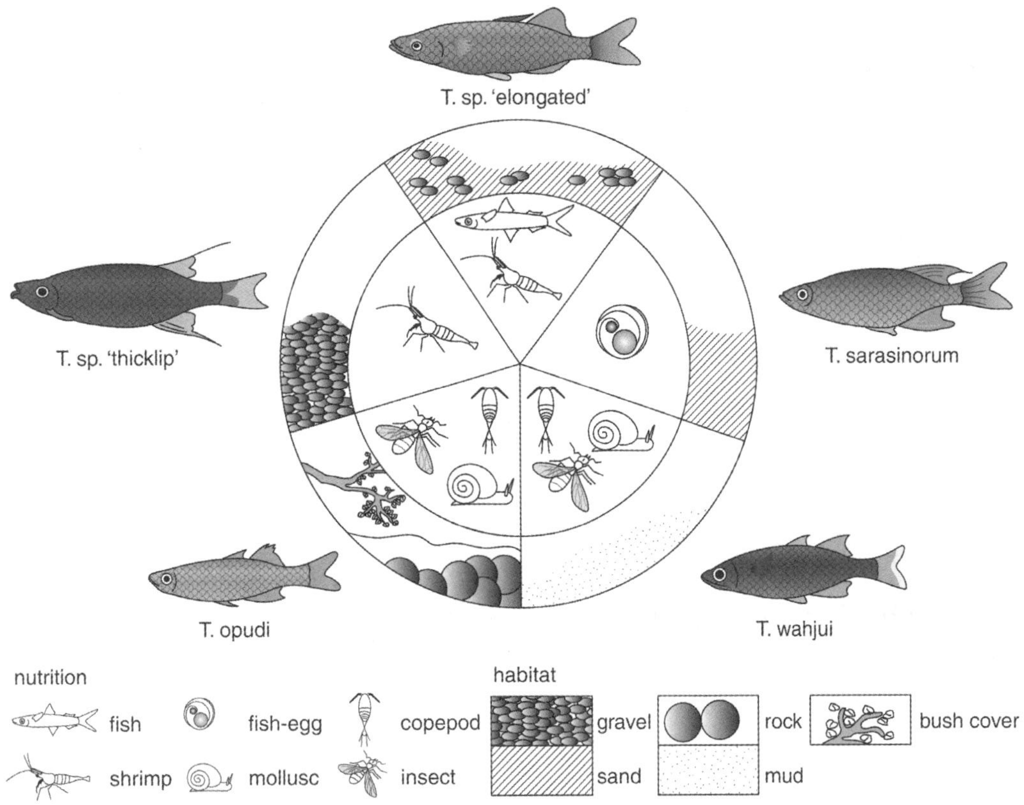

05.1 The copepods in this community are primary consumers.

Suggest what their diet may consist of.

_____

[1 mark]

Question 5 continues on the next page

05.2 In one year, the numbers of *T. sarasinorum* saw a huge increase.

How would this affect the numbers of 'thicklip'?

Explain your answer.

[3 marks]

05.3 Explain why *T. opudi* and *T. wahjui* are **not** competitors, even though they have similar diets.

[2 marks]

05.4 Name a source of pollution that could affect the fish.

[1 mark]

06

06.1 A disease called Leigh syndrome occurs when the process of protein synthesis is disrupted, causing the wrong protein to be made.

Explain how the process of protein synthesis might be disrupted in Leigh syndrome.

[6 marks]

06.2 Give **one** application for our understanding of the human genome.

[1 mark]

Question 6 continues on the next page

06.3 A gardener has been breeding roses in her garden.

She selects the roses with the biggest blossoms and most fragrant flowers to breed together, and pollinates them herself.

A farmer's cabbages suffer from white fly.

The farmer asks a local plant laboratory to create him a resistant breed of cabbage.

Describe and compare the differences between the gardener's and the farmer's approaches to improving their plants.

[4 marks]

07

07.1 Explain the difference between **population size** and **population density**.

[2 marks]

07.2 Mr Green needs to assess the population of plantain on a 10 m wide path in a national park.

Figure 7.1 shows broadleaf plantain, which is a tough plant often found on footpaths.

Figure 7.1

Mr Green has a 25 cm² wire quadrat and a measuring tape.

He places the tape across the path, including the dense verges either side of the path.

What is the name of this line?

[1 mark]

Question 7 continues on the next page

07.3 Mr Green places the quadrat at the end of the line, in the verge.

He counts the number of whole plants in the quadrat and records the number.

How should Mr Green decide where to place the **next** quadrat along the line?

....

....

....

[2 marks]

07.4 Mr Green samples along the line, until he reaches the other end.

The whole path is 500 m long.

Describe the steps Mr Green should follow so that he has statistical evidence for the distribution of plantain **along the length of the path**.

....

....

....

....

[3 marks]

07.5 Explain why there are likely to be more plantains in the **middle** of the path than at the edges.

....

....

[2 marks]

END OF QUESTIONS